BASIC CHEMISTRY

A COMBINED VOLUME

UMESH KUMAR

TECHNIC POINT

Copyright © 2026 TECHNIC POINT

All rights reserved. No part of this publication may be reproduced, distributed, or transmitted in any form or by any means, including photocopying, recording, or other electronic or mechanical methods, without the prior written permission of the publisher, except in the case of brief quotations embodied in critical reviews and certain other noncommercial uses permitted by copyright law.

FOREWORD

In the rapidly evolving landscape of technical education, the challenge is no longer the lack of information, but the overwhelming abundance of it. As an educator who has spent over a decade and a half in the lecture halls and laboratories, I have seen firsthand how students can become lost in technical jargon before they ever grasp the core logic of a circuit or a theory.

This book is the culmination of that experience. It is built on the belief that a teachers primary job is to be a filter to remove the noise and leave only the signal. My journey through Electronic Science has taught me that mastery is found in the fundamentals.

I wrote this volume not just to provide answers, but to foster the kind of analytical thinking required in todays competitive environment. It is my contribution to a generation of learners who seek clarity in a world of complexity.

INTRODUCTION

Necessity is the mother of invention. Human curiosity and imagination had added a wing to our flight of invention. Science has no destination but journey. As a domain of science chemistry also deals with what? Why? How? From what the matter around us made up of? And what are the properties of these matters? These are the subjects of chemistry. It has a golden history from alchemists to the detection of higgs boson (commonly referred by journalists "the god particle"). It is assumed that alchemists used to convert other elements into gold. Now we are observing the same thing only the difference between two elements is the number of fundamental particles electron, proton, and neutrons. Now, we are aware of that the whole universe is made up of 118 elements (discovered till now). Our body, the daily items we use, medicine we take, and our surroundings all are the result of these elements only.

This book can be treated as a general course for secondary level students. It includes all fundamental chapters of chemistry. For convenience of study, it is divided into two parts.

PREFACE

Preface 2014

This book is genuinely written for grasping the fundamental concept of chemistry. It is aimed to the secondary level students. It can serve as a reference for a particular topic. It is also useful for various competitions.

Preface 2026
"The world of science is a journey of constant discovery. When the first edition of Basic Chemistry was released in 2014, my goal was to simplify the fundamental building blocks of matter. Twelve years later, while the core laws remain solid, our interaction with science has transformed. This 2026 Second Edition is meticulously updated to bridge the gap between classic theory and the rapidly evolving requirements of modern competitive exams. From the chemistry of lithium-ion batteries to the environmental impact of carbon compounds, this edition is designed to inspire conceptual clarity and analytical thinking over rote memorization."

PROLOGUE

The transition from theory to practice is often the most difficult leap for any student. In the class room, we see the equations; in the laboratory, we see the components. But there is a silent space in between where true understanding lives.

This book was born from fifteen years of watching students navigate that space. It is designed to strip away the unnecessary complexity that often clouds the beauty of the subject. Whether you are preparing for a competitive examination or seeking to master the fundamentals of the craft, these pages are your roadmap.

Complexity is easy; simplicity is a challenge. My goal is to guide you toward the latter, ensuring that the knowledge you gain here is not just memorized, but mastered.

HOW TO USE THIS BOOK

- **Follow the Two-Part Journey:** Start with **Part 1 (Chapters 0-4)** to master atomic structure and matter before moving to the reactive chemistry and classification in **Part 2 (Chapters 5-9)**.

- **The "Postal Address" Logic:** Throughout the book, use the "Postal Address" method in Chapter 0.4 to visualize electronic configurations. Instead of memorizing codes, think of atoms as buildings with floors (Quantum Numbers) and rooms (Subshells).

- **Deep Dive Challenges:** At the end of every chapter, test your understanding with updated MCQs and analytical problems designed for modern testing trends.

- **Visual Learning Anchors:** Pay close attention to the energy band diagrams and orbital visualizations provided to make abstract concepts like semiconductors tangible.

INTRODUCTION

Necessity is the mother of invention. Human curiosity and imagination had added a wing to our flight of invention. Science has no destination but journey. As a domain of science chemistry also deals with what? Why? How? From what the matter around us made up of? And what are the properties of these matters? These are the subjects of chemistry. It has a golden history from alchemists to the detection of higgs boson (commonly referred by journalists "the god particle"). It is assumed that alchemists used to convert other elements into gold. Now we are observing the same thing only the difference between two elements is the number of fundamental particles electron, proton, and neutrons. Now, we are aware of that the whole universe is made up of 118 elements (discovered till now). Our body, the daily items we use, medicine we take, and our surroundings all are the result of these elements only.

This book can be treated as a general course for secondary level students. It includes all fundamental chapters of chemistry. For convenience of study, it is divided into two parts.

Part 1 includes chapter pertaining to some elementary concepts of chemistry, structure of atom, atom and molecules and matter around us.

Part 2 includes chemical reactions and equations, acid, bases and salt, metals and nonmetals, carbon compounds and the last lesson deal with periodic classification of element.

PART 1

CHAPTER 0

SOME ELEMENTARY CONCEPTS

0.1 ELEMENTS

ELEMENTS: According to the Indian Vedas the whole universe is made up of five fundamental sources (building blocks). These are soil, water, fire, sky and air. As per our curiosity and experiment, we verified that these are not ultimate sources. Water is the combination of hydrogen and oxygen. Air is a mixture of many gases like nitrogen, hydrogen, chlorine etc. These basic building blocks are referred as elements. Till now we have 94 natural occurring elements and 24 synthesized in labs.

Atomic no.(z)	Name of the elements	Symbols	Atomic Mass Unit (amu)
1	Hydrogen	H	1
2	Helium	He	4
3	Lithium	Li	7
4	Beryllium	Be	8
5	Boron	B	11
6	Carbon	C	12
7	Nitrogen	N	14
8	Oxygen	O	16
9	Fluorine	F	19
10	Neon	Ne	20
11	Sodium	Na	23
12	Magnesium	Mg	24
13	Aluminum	Al	27
14	Silicon	Si	28
15	Phosphorus	P	31
16	Sulphur	S	32
17	Chlorine	Cl	35
18	Argon	Ar	40
19	Potassium	K	40
20	Calcium	Ca	40

First 20 elements with their atomic no., atomic weight, and symbol

Some other elements and their symbols which are derived from their Latin name

Element (English)	Latin Name	Symbol
Sodium	Natrium	Na
Potassium	Kalium	K
Iron	Ferrum	Fe
Copper	Cuprum	Cu
Silver	Argentum	Ag
Gold	Aurum	Au
Mercury	Hydrargyrum	Hg
Lead	Plumbum	Pb
Tin	Stannum	Sn
Antimony	Stibium	Sb

0.2 ATOM

ATOM: the word "atom" is derived from the Latin word atomos which means uncut able. In India around 500 B.C maharishi Kanad gave the word parmanu which is now referred as atom. Each element has its unique atom from which it is made up of.

0.3 ELECTRONIC CONFIGURATION

ELECTRONIC CONFIGURATION: Each atom has nucleus which consists of neutron and proton. Electrons revolve around the nucleus in specific elliptical orbits. Electronic configuration is the postal address of the electrons in an atom. In chemistry, electronic configuration is the basic tool through which we can find the desired property of an element. We can find the states of the element, valency of the atom. These are helpful in choosing a particular element for a particular purpose.

HOW TO FIND VALENCY FROM ELECTRONIC CONFIGURATION:

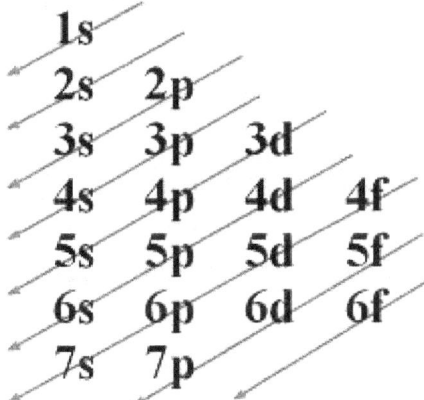

Fig.1: Filling of electrons in orbits

We can find the valency of a particular element in simple steps. Before using these steps, we must know that the filling of electron is started from 1s then 2s and go on succession with 2p, 3s, 3p, 4s, 3d, 4p, 5s, 4d, 5p, 6s, 4f, 5d, 6p, 7s, 5f, 6d, 7p, 6f and so on as shown in fig. 1. Secondly, we must also care that the max no. of electron in a particular shell-like s, p, d, and f is limited i.e.

Shell	Maximum no. of electron allowed
s	2
p	6
d	10
f	14

Working step:

Example 0.1: find the valency of sodium? Solution:
Step: 1 find the electronic configuration

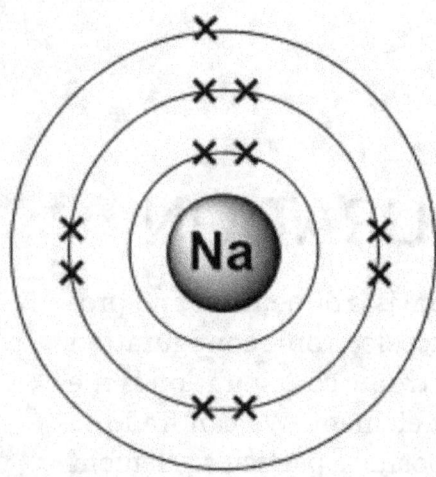

The atomic no. of sodium is 11. So, there are 11 electrons in its neutral atom. Therefore, filling of electrons will be as follows $1s^2, 2s^2, 2p^6, 3s^1$.

Fig.2 electronic configuration of Na

Note: here we find that the max no. of electron can be 2 in 3s shells but already 10 electrons have been filled in 1s, 2s and 2p, so only 1 electron is filled in 3s.

Step: 2 in the outer most shell there is only one electron. Hence its valency is one.

Note: valence electron refers to the no. of electrons residing in the outer most shell and only these electrons take part in a particular reaction. If there are more than 4 electrons in the outer most orbit. For finding valency we must subtract the no. of electrons from 8 (as it is necessary to complete octet). But this time valency is in negative due to receiving electron from other atom electron has negative charge.

We can see it in the next example:

Example 0.2: find the valency of oxygen? Solution:

Step: 1 find the electronic configuration

The atomic no. of oxygen is 8. So, there are 8 electrons in its neutral atom. Therefore, filling of electrons will be as follows $1s^2, 2s^2, 2p^4$.

Note: Here we find that the max no. of electron can be 6 in 2p shell but already 4 electrons have been filled in 1s and 2s so only 4 electrons is filled in 2p.

Step 2: In the outer most shell there are 6 electrons (2 in 2s and 4 in 2p as outer most orbit is 2) only, which is greater than 4. Hence its valency 6-8=-2.

BASIC CHEMISTRY

ELECTRONIC CONFIGURATION OF OXYGEN
(O, Atomic Number 8)

A. THE OXYGEN ATOM: Core and Valence Electrons
- Nucleus (8p, 8n)
- 1st Shell (Core): 2 e⁻ ($1s^2$)
- 2nd Shell (Valence): 6 e⁻ ($2s^2\ 2p^4$)
- Total Electrons = 8

B. AUFBAU DIAGRAM / ENERGY LEVEL SCHEMATIC

Increasing Energy:
- 2p: $2p_x$ ↑↓, $2p_y$ ↑, $2p_z$ ↑
- 2s: ↑↓
- 1s: ↑↓

C. ORBITAL BOX DIAGRAM (Hund's Rule and Pauli Exclusion)

1s [↑↓] → 2s [↑↓] → 2p [↑↓][↑][↑][ ]

D. 3D VISUALIZATION OF ORBITALS

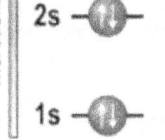

Labels: 1s orbital, 2s orbital, $2p_x$ orbital, $2p_y$ orbital, $2p_z$ orbital

E. ELECTRONIC CONFIGURATION NOTATION

Standard: $1s^2\ 2s^2\ 2p^4$ — 4 valence electrons in 2p

Noble Gas (He): $[He]\ 2s^2\ 2p^4$ — He Core ($1s^2$)

F. KEY TAKEAWAYS
* Valence Electrons: 6.
* Core Electrons: 2.
* Full Subshells: 1s, 2s.
* Half-full Subshells: None
* Unpaired Electrons: 2 (in 2p).
* Total Spin: 1.

0.4 THE "POSTAL ADDRESS" METHOD FOR ELECTRONIC CONFIGURATION

When scientists write the electronic configuration of an atom (e.g., $1s^2\ 2s^2\ 2p^1$ for Boron), they aren't just writing random codes. They are providing the exact **postal address** for every electron in that atom.

Most students get confused because they try to memorize the order (1s, 2s, 2p, 3s). Stop memorizing. Start visualizing.

Think of an atom as a high-rise building with several **Floors** (this is the Principle Quantum Number, n). On each floor, there are different types of **Rooms** (these are the Subshells, l: s, p, d, f). Each room type has a strictly fixed capacity for **Residents** (these are the Electrons).

Refer to the diagram below to understand this relationship:

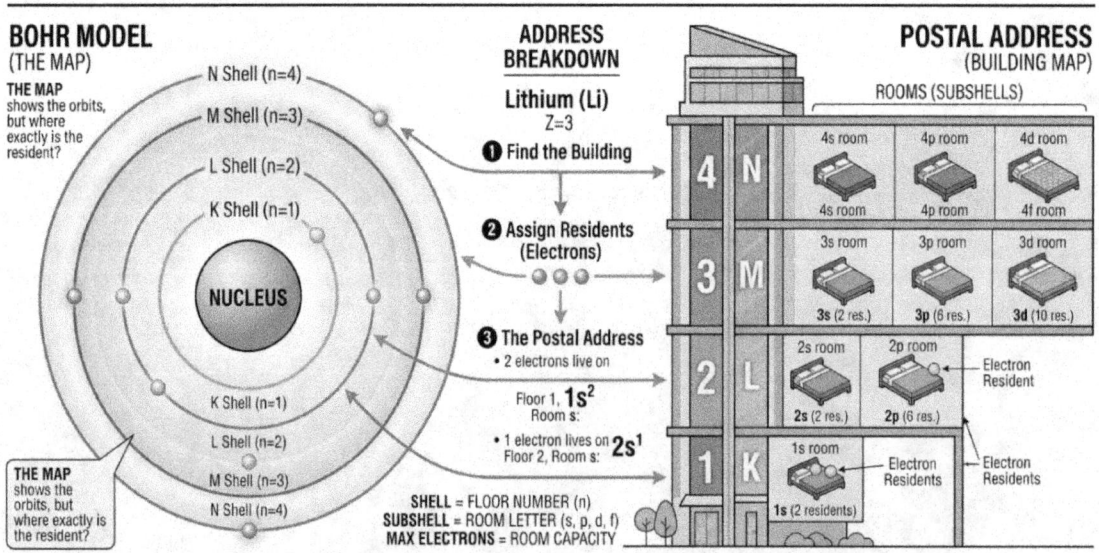

As the diagram shows:
- Bohr Shells (K, L, M) → Building Floors ($n = 1,2,3$).
- Subshells (s, p, d, f) → Room Types on that Floor
- Electrons → The Residents.
  o An s-room always holds 2 residents.
  o A p-room always holds 6 residents.

o A d-room always holds 10 residents.

The address always follows this strict rule: **Floor Number + Room Letter + Resident Count (written as a superscript).**

0.5 CHEMICAL REACTIONS USING VALENCY

Why does reaction take place?

It is general question, why does any reaction take place? When H_2 and O_2 combine only H_2O is formed not other compound? Why does H_2 Combine with O_2 but not with He? Every atom wants to get a stable configuration. Any atom having full filled outer orbit is considered as a stable atom. An atom can have 8 electrons in its outer most orbits to get its stable condition referred as octet. In case of H
and He, it is 2 and referred as duplet. When two or more atoms come in vicinity or contact, they exchange their electron so as to get a stable configuration. The combining capacity of an atom is referred as valency.

For example: when Na and Cl react, Sodium has 1 electron in its outer orbit and chlorine has 7. If Na loses its 1 electron and chlorine gains this both will have a stable configuration. In this process Na changed to +ve ion and Chlorine to –ve. Now ionic bonding makes it possible to form NaCl.

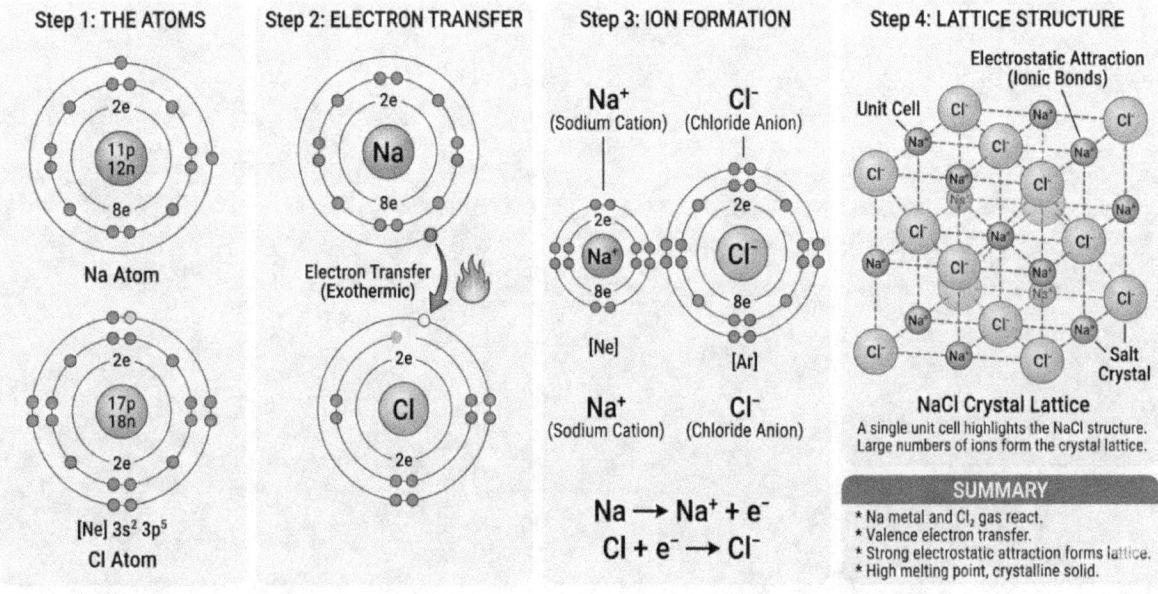

In the above example we see that the actual process is the swapping or exchange of valencies. If we know the valences' we can easily write the chemical formula.

Following examples will illustrate the concept:

1. Carbon Disulfide (CS_2) Carbon belongs to Group 14 and has 4 valence electrons, while Sulfur belongs to Group 16 and has a valency of 2 (since it needs 2 electrons to complete its octet).
   - Symbols: C and S
   - Valencies: $C = 4$, $S = 2$
   - Criss-Cross: The 4 from Carbon goes to Sulfur, and the 2 from Sulphur goes to Carbon.
   - Initial Formula: C_2S_4
   - Simplified: We divide by the greatest common divisor (2) to get CS_2.

2. Aluminum Oxide (Al_2O_3) Aluminum is a metal with a valency of 3, and Oxygen is a non-metal with a valency of 2.
   - Symbols: Al and O
   - Valencies: $Al = 3$, $O = 2$
   - Criss-Cross: The 3 from Aluminum moves to Oxygen, and the 2 from Oxygen moves to Aluminum.
   - Final Formula: Since 2 and 3 cannot be simplified further, the formula is Al_2O_3.

3. Hydrogen Fluoride (HF) Hydrogen typically has a valency of 1, and Fluorine (a halogen) also has a valency of 1.
   - Symbols: H and F
   - Valencies: $H = 1$, $F = 1$
   - Criss-Cross: Both elements exchange the number 1.
   - Final Formula: In chemistry, the subscript "1" is not written. Therefore, the formula is HF

FORMATION OF CS$_2$ (CARBON DISULFIDE)   FORMATION OF HF (HYDROGEN FLUORIDE)

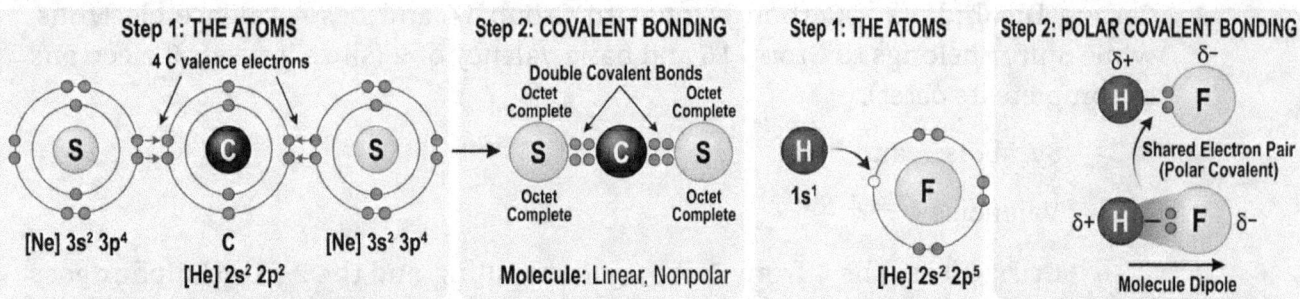

FORMATION OF Al$_2$O$_3$ (ALUMINUM OXIDE)

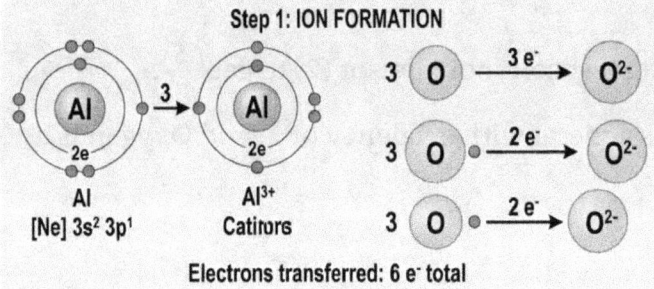

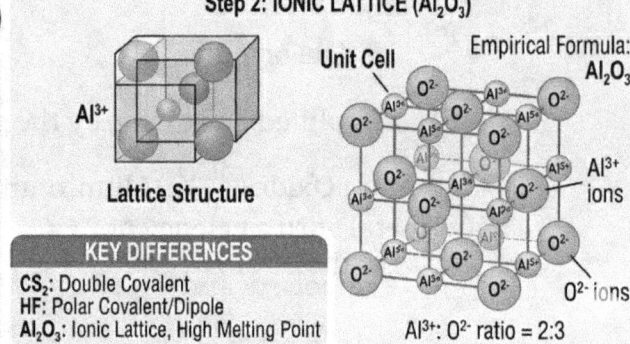

KEY DIFFERENCES
CS$_2$: Double Covalent
HF: Polar Covalent/Dipole
Al$_2$O$_3$: Ionic Lattice, High Melting Point

CHAPTER 0: DEEP DIVE CHALLENGES

Q1. Nucleus consists of?

   (a) Electron and neutron

   (b) Electron and proton

   (c) Proton and neutron

   (d) None of these

Q2. Valency of oxygen is?

   (a) 2

   (b) -2

   (c) 1

   (d) None of these

Q3. $Na + Cl_2 \rightarrow$? What is the product of above reaction?

   (a) $NaCl_3$

   (b) $NaCl_2$

   (c) $NaCl$

   (d) None of these

Q4. How many elements has synthesized in lab till now?

   (a) 92

   (b) 26

   (c) 24

   (d) 118

Q5. Which is an inert gas?

(a) Hydrogen

(b) Nitrogen

(c) Oxygen

(d) Neon

Q6. Which one is the heaviest among these?

(a) Electron

(b) Proton

(c) Neutron

(d) All has same masses

Q7. Which one has independent existence in a chemical reaction?

(a) Electron

(b) Atom

(c) Molecule

(d) None of these

Q8. How many electrons are there in a Na+ ion?

(a) 10

(b) 11

(c) 12

(d) 9

Q9. Why does any atom react with other?

(a) to give its electron

(b) to gain electrons

(c) to get a stable configuration.

(d) none of these

Q10. Cl_2 is?

(a) Monatomic

(b) diatomic

(c) Inert gas

(d) none of these

11. Choose whether these statements are true or false

   a) Electrons have negative charge.

   b) In a simple hydrogen atom one electron and one proton is present.

   c) Electrons revolve around the nucleus.

   d) Electronic configuration is helpful in finding whether a given element is inert or not.

   e) Chlorine has positive valency.

   f) Hydrogen molecule is diatomic

SOLUTION

1. C 2. B 3. C 4. C 5. D 6. C 7. C 8. A 9. C 10. B

11. TRUE A, B, C, D, F, FALSE E

CHAPTER 1
STRUCTURE OF ATOM

1.1 DISCOVERY OF ELECTRON

What are insulators and conductors?

The materials which allow electricity to pass through them are known as conductor and those not known as insulator.

We can demonstrate it in a simple way. We break the wire of a simple closed circuit containing a DC battery, bulb and connecting wire. Then join the material ends to the ends of wires. If bulb glows it's a conductor otherwise insulator.

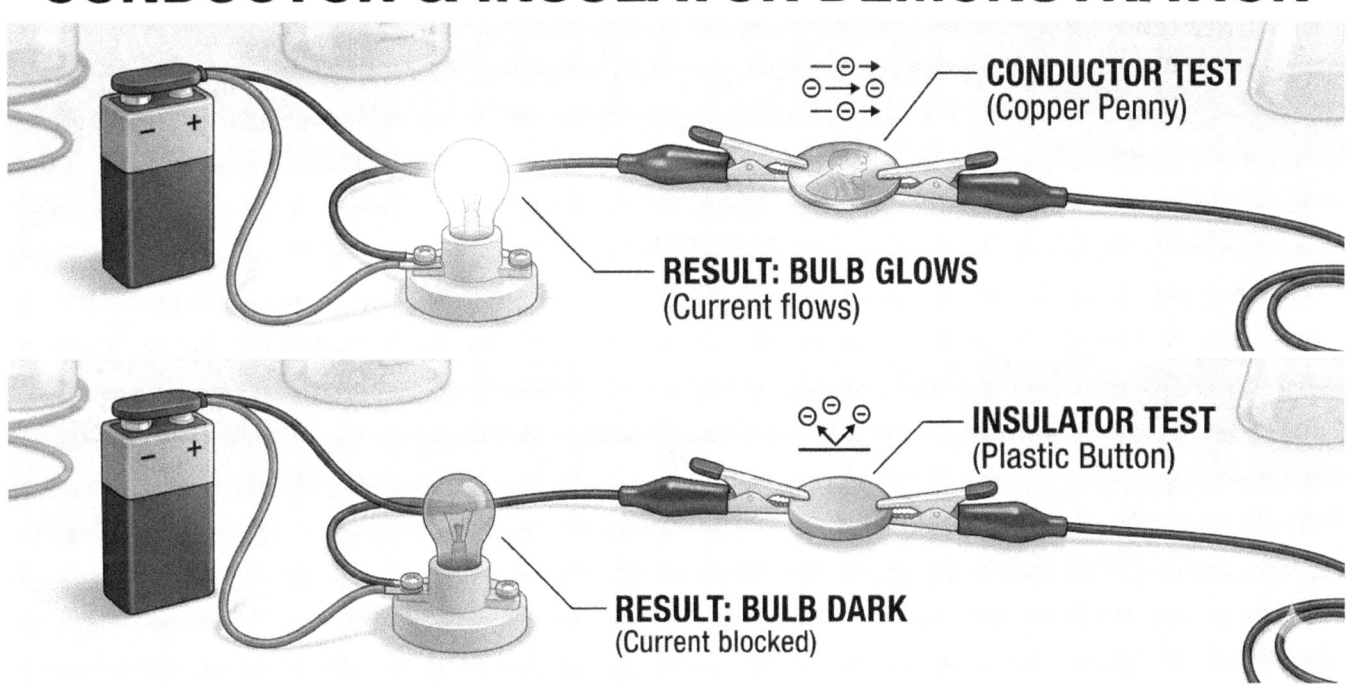

How to demonstrate whether a liquid is insulator or not?

A liquid can be also demonstrated in the similar manner as stated above. Only slight change is that the liquid is kept in a beaker or insulated pot. The open ends of the wire are kept in the liquid and glowing of bulb is observed.

Is air insulator? How to demonstrate it?

If we take in the succession of above experiment, here we have to take a closed tube for gas or air taken under observation. In general, it seems that air is insulator at normal temperature and pressure. It is right. Because if it was a conductor at room temp. and pressure our room had electrified by switch of electric board and we would not have survived as now.

But what at a very high voltage and low pressure supplied to the taken discharge tube. J.J Thomson raised the voltage more than 10000v in the similar experiment and reduces the pressure introducing a vacuum tube in the discharge tube. He observed the flickering yellow greenish light on the ZnS plate kept opposite of the cathode.

It was concluded that the air is conductor at very high voltage and extremely low pressure. One important conclusion from this experiment leads to the discovery of electron. The ray which illuminated the ZnS screen was termed as the cathode ray as it was originated from cathode. It was composed of the negatively charged particle termed as **electron.**

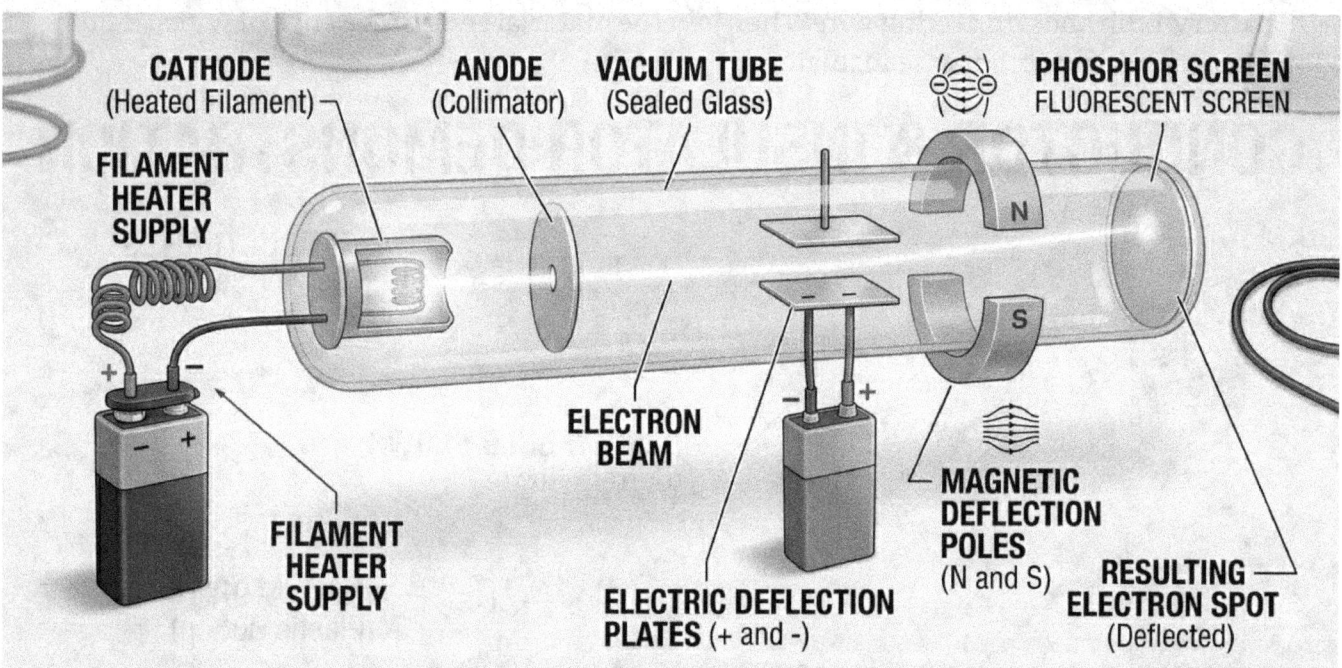

1.2 DISCOVERY OF PROTON

At that time, it was assumed that the atom as a whole is neutral. After the discovery of electron it was seen that on removal of electron the rest part must be positively charged. For this perforated cathode was taken and the screen was kept opposite the anode. The remaining +ve charged where moved towards the screen and seen as a fluorescent on the screen. They were passing through hole or canal of the cathode and termed as the **canal ray**. The constituent particle of the canal ray was examined and given the name proton.

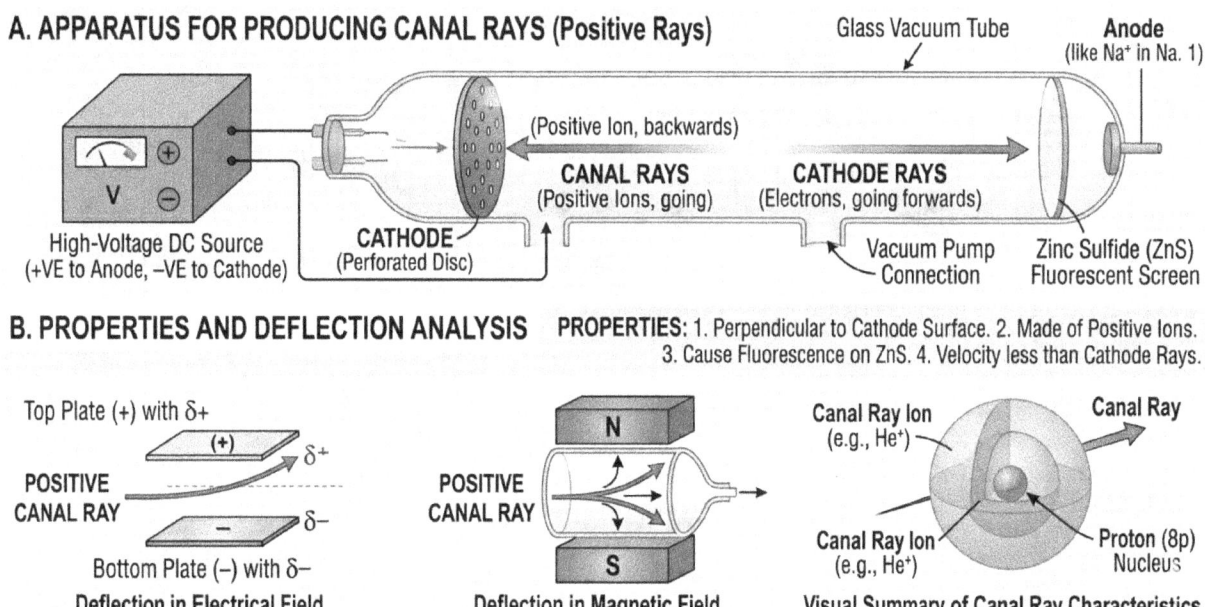

1.3 DISCOVERY OF NEUTRON

It was observed that the mass of atom other than electron was approximately double of the proton. So, there must be a particle having mass equal to the proton and neutral in charge inside the atom. The name of this particle was given by JAMES CHADWICK. It is now referred as neutron.

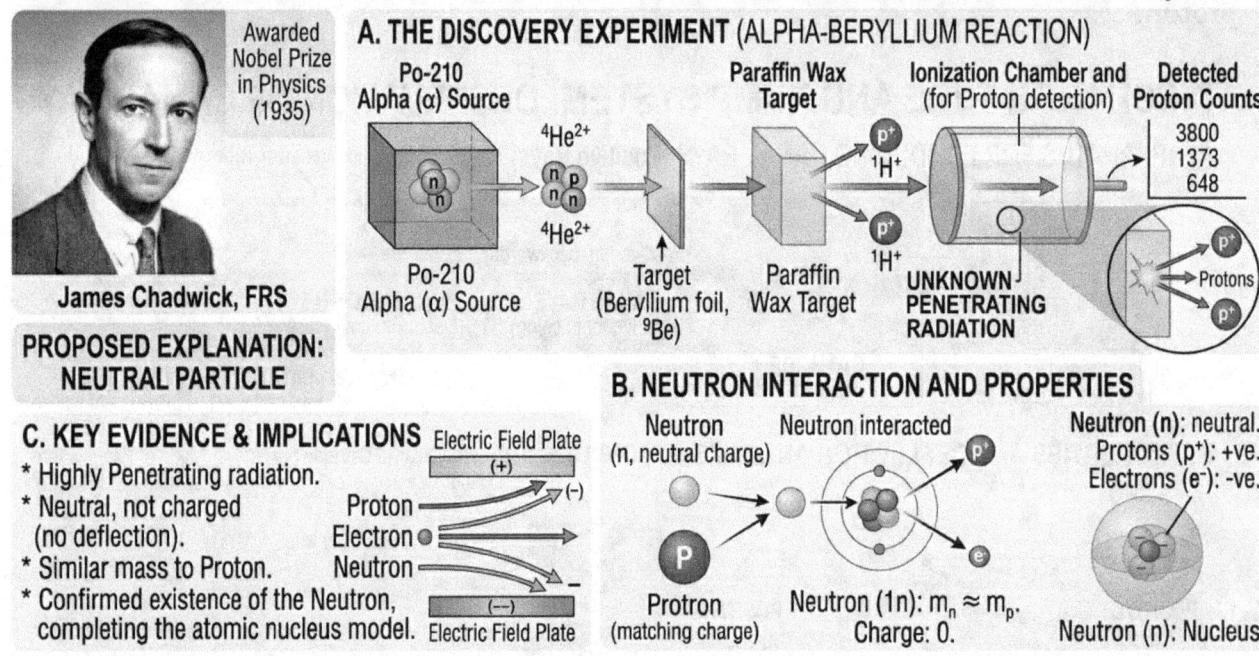

1.4 ATOMIC MODELS

After the discovery of electron, proton and neutron, it was a big challenge to represent the arrangement of these sub atomic particles inside the atom. First model was forwarded by J.J Thomson.

1.5 J. J THOMSON MODEL OF ATOM

J.J Thomson gave the plum pudding modeling. According to him the atom was like a watermelon. The seeds were referred as the electrons embedded in the positively charged watermelon. The +ve and –ve charge were balanced to provide a neutral atom.

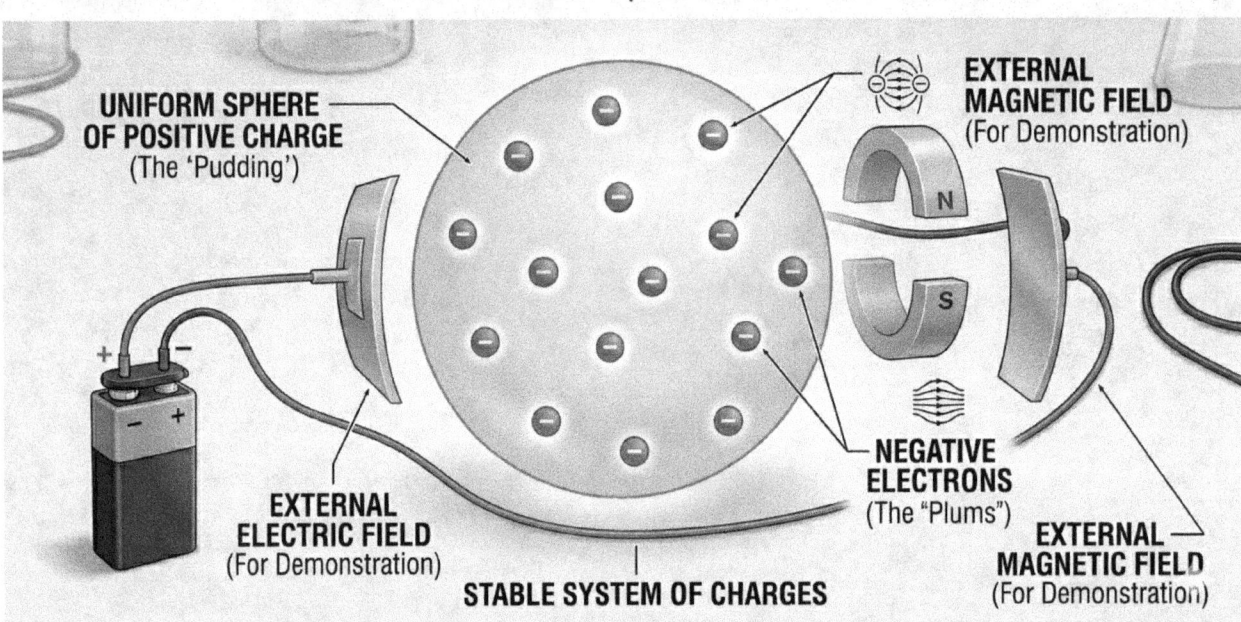

1.6 RUTHERFORD'S SCATTERING EXPERIMENT

Aim: earnest Rutherford was interested in finding the arrangement of the electrons inside the atom. For satisfying his curiosity he conducted an experiment.

Arrangement: He was interested in taking a very thin layer of atom. For this purpose, he took a fine layer of gold. It was approximately 100 atoms thick. This thin layer was bombarded with alpha particle.

Observation:

a) Almost all alpha particles were going without any deviation.

b) Very few were deflected at small angle.

c) A single particle in 12000 approximately deflected at 180 degrees.

Conclusion:

a) Most of the part inside the atom is empty.

b) As alpha particle is positively charged and approximately 8000 times heavier than electron but it is returning so almost all mass of atom and positively charges must reside at its centre.

c) Almost 99% alpha particle is going undeviated so size of nucleus is very small.

1.7 RUTHERFORD'S ATOMIC MODEL

As per the experimental result Rutherford gave the following atomic model

a) Atom is like a sphere and there is nucleus in its centre. Nucleus is positively charged and whole mass
    of atom resides inside nucleus.

b) Electrons revolve round the nucleus in circular orbit.

c) The size of nucleus is very small.

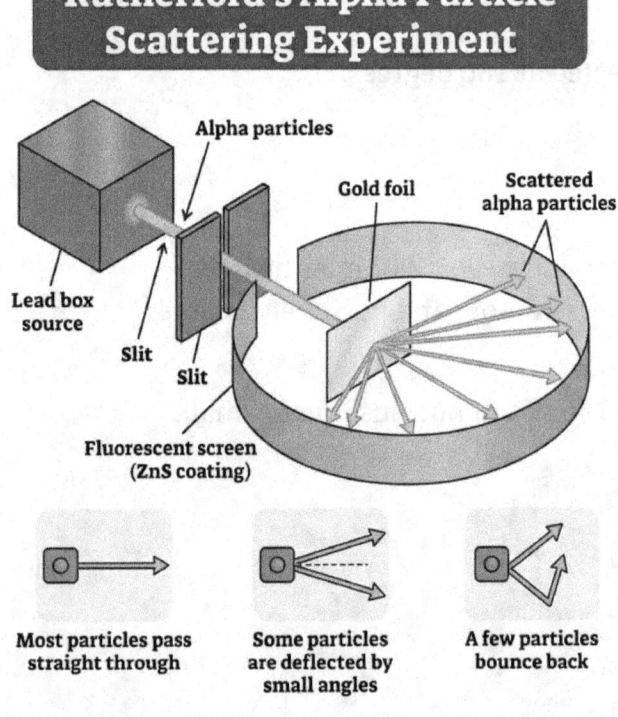

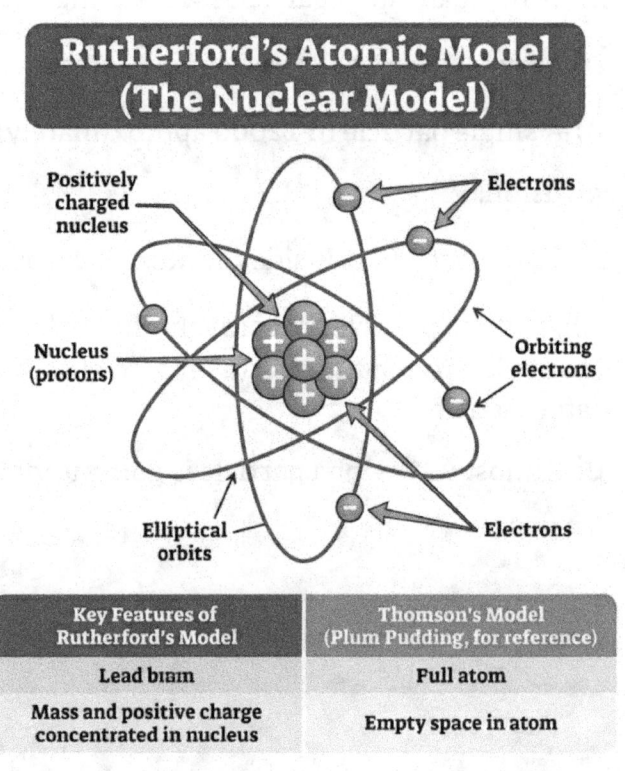

1.8 DRAWBACKS OF RUTHERFORD'S ATOMIC MODEL

According to the Rutherford's atomic model, electron revolves around the nucleus. As per electromagnetic theory, every moving object emits electromagnetic radiations and thus loses its energy. It is just like running in a circular park. If so, happens electron must lose energy and spiral into the nucleus within a time span of 10^{-6} second. In this case electron will not exist so atom and so, this universe, but it is! Rutherford was unable to explain it. It was a great drawback of Rutherford's atomic model.

Rutherford's Atomic Model (1911): A Summary

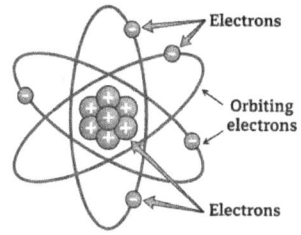

Key features of Rutherford's model:
- Positvet,nd positively charged nucleus
- Mass and positivię charge concentrated in nucleus with electrons → shrinking orbit

Key Atom Collapse
- Maxwelll predicts an accelerating charged electron must radiate energy
- Massesits orbit to orbit in atom modelᴛ! increasetᴛ positively Rutherford into nucleus

Major Drawback: Atom Stability (Maxwell's Electromagnetic Theory)

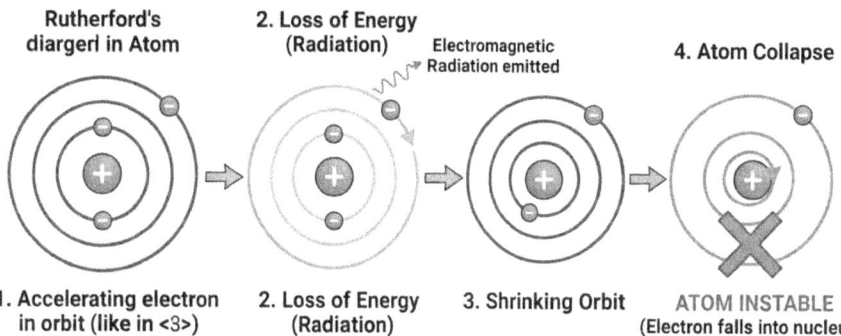

1. Accelerating electron in orbit (like in <3>)
2. Loss of Energy (Radiation)
3. Shrinking Orbit
4. ATOM INSTABLE (Electron falls into nucleus)

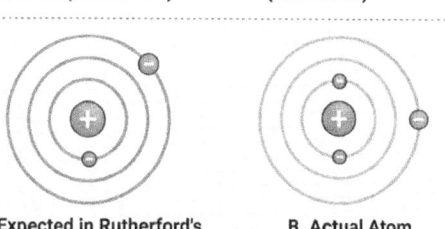

A. Expected in Rutherford's Model (Spiral, Instable)
B. Actual Atom (Stable Orbit, observed)

Maxwell's theory trute predicts that an accelerating charged electron must radiate energy. Wihin o raacurre dense energy, in the tratvel, causing its orbit to shrinks, the atom data in nucleus. The atom is collapse within a IPTs fraction of a second.

1.9 BOHR'S MODEL OF ATOM

Bohr's model resolves the drawback of Rutherford's atomic model. According to this model

a) Every electron moves in a fixed energy orbit.

b) They do not lose energy during their revolution.

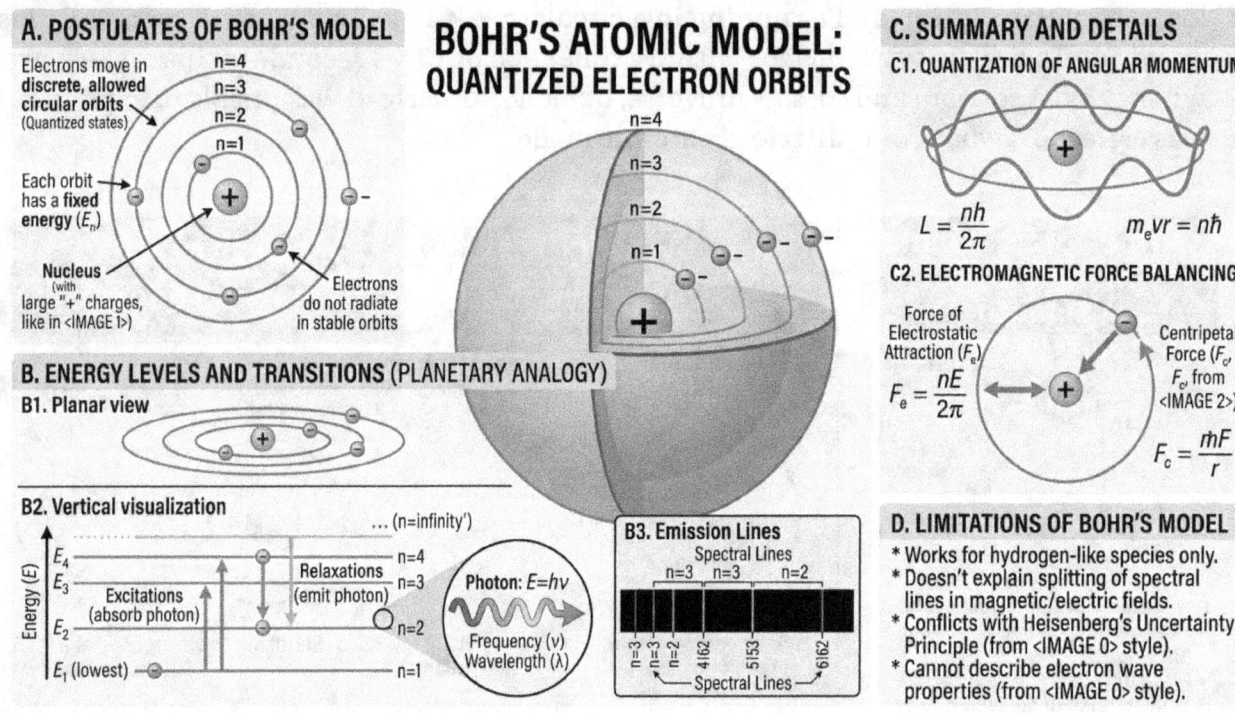

1.10 SOMERFIELD'S ATOMIC MODEL

Somerfield explained that the electrons do not revolve in a circular orbit but elliptical.

SIMPLE EXPLANATION OF SOMMERFELD'S ATOMIC MODEL

A. Comparing Orbits with Bohr Model

1. BOHR MODEL (from <Image 1>)

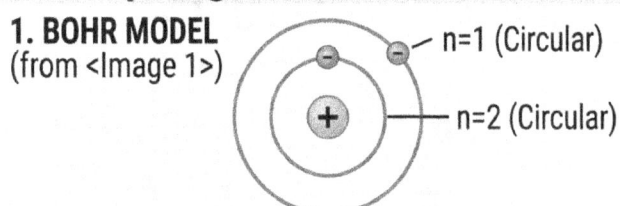

- n=1 (Circular)
- n=2 (Circular)

2. SOMMERFELD MODEL (Introduction of Ellipses)

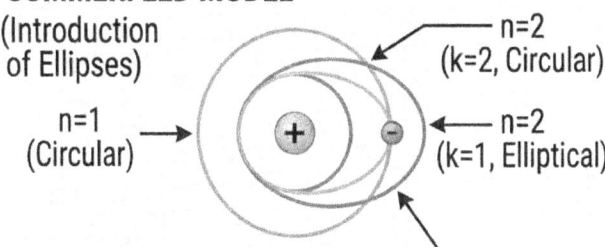

- n=1 (Circular)
- n=2 (k=2, Circular)
- n=2 (k=1, Elliptical)

Key Idea: Orbits are Elliptical in general. Multiple shapes possible for the same principal number "n".

Electron speed changes in elliptical path.

B. How it Resolves Fine Structure

1. SPECTRUM IN BOHR MODEL

| n=2 to n=1 transition | **Result:** A single, "fuzzy" spectral line. |

2. SPECTRUM IN SOMMERFELD MODEL

n=2 (Circular) to n=1 ⟶ | ⟵ n=2 (Elliptical) to n=1

The energy is slightly different. The result is:
SPLITTING OF SPECTRAL LINES (Fine Structure).
• Explains "why" spectral lines seem to split.

KEY REFINEMENTS OVER BOHR MODEL	
Bohr	**Sommerfeld**
Orbits: Circular only	**Orbits:** Elliptical & Circular
Quantum Numbers: 'n' only	**Quantum Numbers:** 'n' & 'k' (shape)
Spectral Result: Single line	**Spectral Result:** Splitting (Fine Structure)

1.11 SIZE OF ATOM AND NUCLEUS

In general form if we consider the nucleus as tip point of our pen's nib. The dimension of our room will be the atom. If we take a cricket ball as a nucleus then a circle of radius 5km will constitute the first orbit of the atom, taken ball as a Centre. If we take nucleus as a Centre of our solar system then in proportion our solar system will be smaller than the atom. That's why now a day it is an area of research and emergence of nanotechnology.

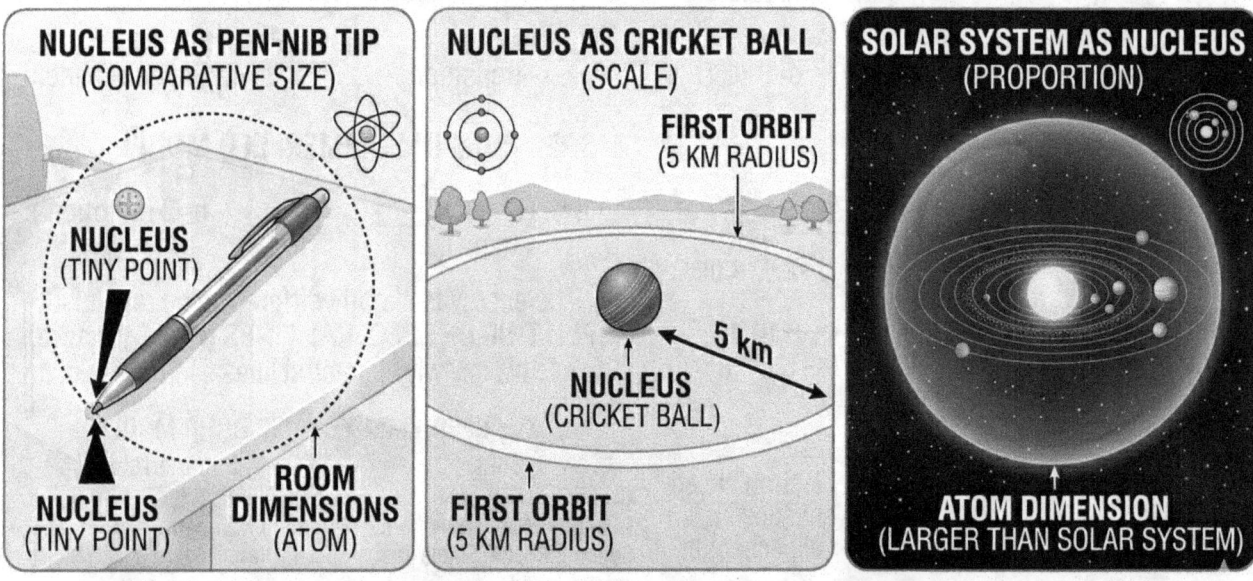

ATOMIC & NUCLEAR SCALE COMPARISONS

EMERGENCE OF NANOTECHNOLOGY AND MODERN ATOMIC RESEARCH

1.12 ISOTOPE

There are many atoms found naturally which have same atomic no. but different atomic masses. These atoms are referred as the isotope of each other.

1.13 ISOBAR

There are many atoms which have same no. of atomic masses but different atomic number. These atoms are referred as isobar of each other.

1.14 ISOTONE

Atoms having same number of neutrons are known as isotones of each other.

1.15 ISODIAPHERS

Isodiaphers are atoms of different elements that have the same neutron excess (the difference between neutrons and protons).
- The Rule: $N - Z$ is constant.
- Example: Uranium-238 ($^{238}_{92}U$) and Thorium-234 ($^{234}_{90}Th$).
  For U: $N = 146, Z = 92 \rightarrow 146 - 92 = \mathbf{54}$
  For Th: $N = 144, Z = 90 \rightarrow 144 - 90 = \mathbf{54}$

1.16. ISOSTERS

These are molecules (not just atoms) that have a similar "shape" and electron count.
- **The Rule:** Same number of **atoms** and same number of **valence electrons**.
- **Example:** Carbon dioxide (CO_2) and Nitrous oxide (N_2O). Both have 3 atoms and 22 valence electrons.

1.17. NUCLEAR ISOMERS

These are the same atom, but with different energy levels.
- **The Rule:** Same Z and same A, but different **energy states** and half-lives.
- **Example:** Technetium-99 (^{99}Tc) and its metastable state Technetium-99m (^{99m}Tc).

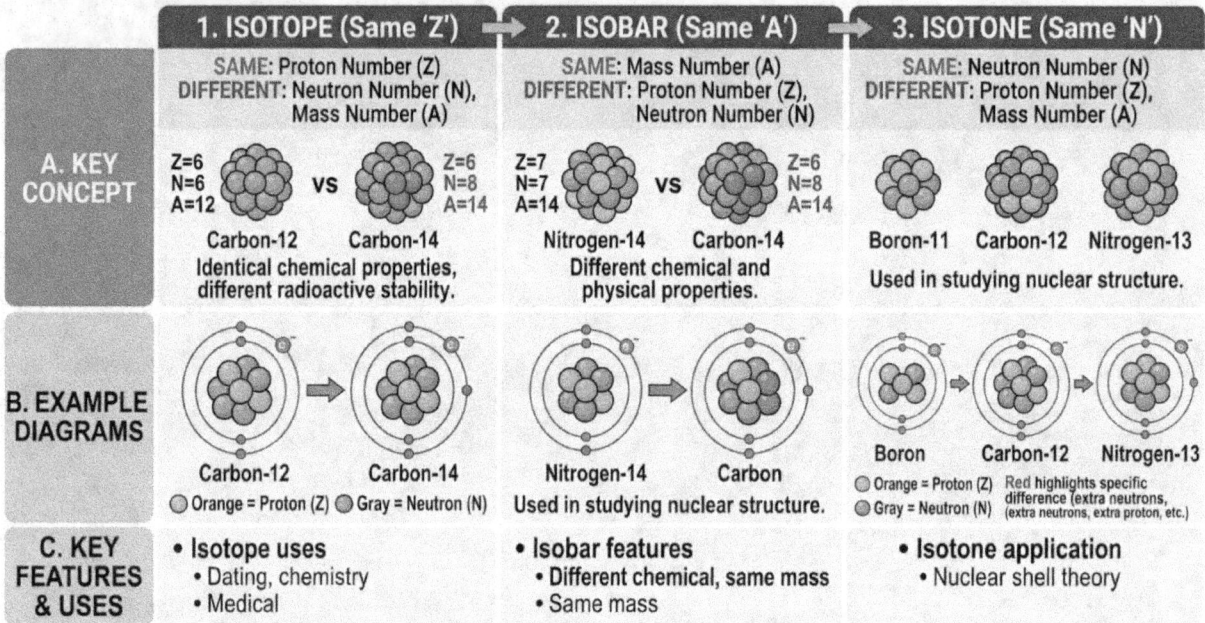

1.18 CONDUCTOR INSULATOR AND SEMICONDUCTOR: ATOMIC CONTEXT

Conductor (Waterfall Mist): On the left, a waterfall creates a dense mist of free-floating electrons, which easily flow in one direction under a "Potential Provided" force.

Insulator (Dry Bedrock): In the middle, a barren canyon with no moisture illustrates how electrons are locked within the atomic structure, preventing any flow even with potential applied.

Semiconductor (Melting Chocolate): On the right, temperature plays a key role. It shows cold, solid chocolate on wafers at the top, and warming, flowing chocolate at the bottom, demonstrating how increased heat releases electrons to flow.

Each section includes a visual diagram of the energy bands to show the scientific mechanism behind the metaphor.

CONCEPTUAL ANALOGIES FOR ELECTRICAL MATERIALS

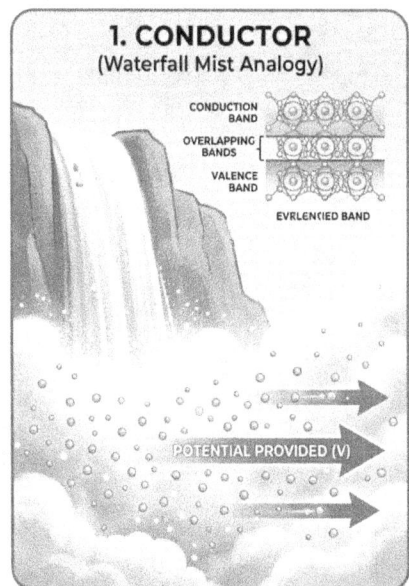

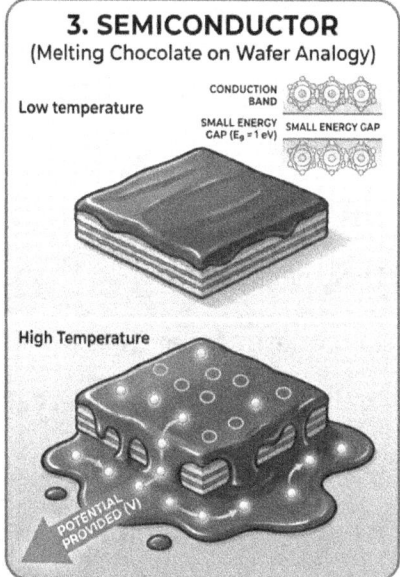

CHAPTER 1: DEEP DIVE CHALLENGES

Q1. Cathode ray consists of?

a) Electron

b) Atom

c) Molecule

d) None of these

Q2. canal ray consists of?

a) Electron

b) Atom

c) Molecule

d) None of these

Q3. Neutron has   charges?

a) Negative

b) Positive

c) None of these

d) Sometime negative some time positive

Q4. When cathode ray observed in J.J Thomson's experiment?

a) At high temperature and high pressure.

b) At very high voltage and very low pressure.

c) At high voltage and high pressure.

d) None of these

Q5. Who discovered electron?

(a) J.J Thomson's

(b) E. Goldstein

(c) James Chadwick

(d) Neils Bohr

Q6. Who discovered neutron?

a) Rutherford

b) E. Goldstein

c) James Chadwick

d) Neils Bohr

Q7. Isotopes have same… ?

a) Atomic masses

b) no. of neutrons

c) Atomic number

d) none of these

Q8. Cl-35/17 and Cl-37/17 are?

a) Isotopes

b) Isobars

c) Isotones

d) None of these

Q9. N-14/7 and C-14/6 are?

a) Isotopes

b) Isobars

c) Isotones

d) None of these

Q10. In alpha particle scattering experiment Rutherford chosen?

a) Zn foils

b) Silver foil

c) Gold foil

d) None of these

Q11. State whether these statements are true or false

a) Rutherford gave the plum pudding model of atom.

b) Somerfield explained that the electrons revolve in an elliptical orbit.

c) Isotones have same no. of protons.

d) The whole mass of atom is concentrated in the nucleus.

e) James Chadwick discovered the neutron.

f) Nucleus is neutral.

g) Atom is neutral but nucleus is negatively charged.

h) Isotope of iodine is used to cure goiter.

Short answer type questions:

Q1. Write two differences between canal ray and cathode ray.

Q2. What was the result of Rutherford's scattering experiment? Mention any two.

Q3. what are isotopes. Write three isotopes of hydrogen.

Long answer type questions:

Q1. Briefly explain the Thomson's model of atom.

Q2. Explain the alpha particle scattering experiment keeping in view following topics:

a) Aim   b) Observation   c) Conclusion

Solution

1.a 2. d 3.c 4. b 5.a 6.c 7.c 8. a 9.b 10.c

11. TRUE b, d, e, h

FALSE a, c, f, g

Short Answer

Q1. Differences between Canal Rays and Cathode Rays
- **Nature of Particles:** Cathode rays consist of negatively charged electrons, while Canal rays (Anode rays) consist of positively charged ions.
- **Mass and Charge:** The properties of cathode rays are independent of the gas in the tube, whereas the properties of canal rays depend on the nature of the gas used.

Q2. Results of Rutherford's Scattering Experiment
- **Nuclear Presence:** Most of the space inside an atom is empty because most alpha particles passed through the gold foil without deflection.
- **Positive Center:** The atom contains a very small, dense, and positively charged center called the nucleus.

Q3. Isotopes and Hydrogen Isotopes
- **Definition:** Isotopes are atoms of the same element that have the same atomic number but different mass numbers.
- **Hydrogen Isotopes:**
    1. **Protium** ($_1H^1$) - No neutrons.
    2. **Deuterium** ($_1H^2$) - One neutron.
    3. **Tritium** ($_1H^3$) - Two neutrons.

Long Answer

Q1. Thomson's Model of Atom
- **Plum Pudding Analogy:** The atom is visualized as a sphere of positive charge with negatively charged electrons embedded in it, like raisins in a pudding.
- **Electrical Neutrality:** The total magnitude of the positive charge (the sphere) is equal to the total magnitude of the negative charge (the electrons), making the atom electrically neutral as a whole.

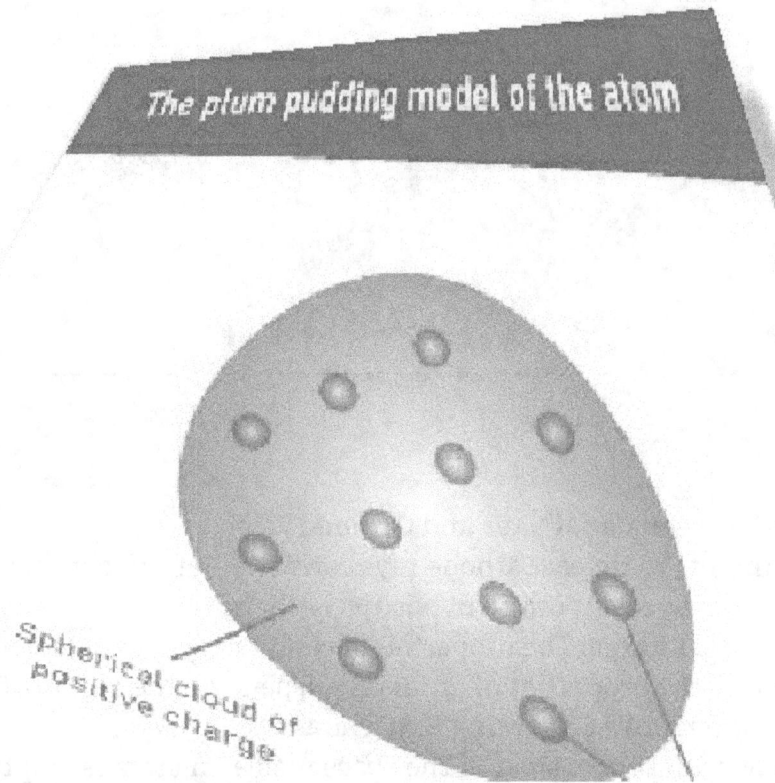

Q2. Alpha Particle Scattering Experiment

- **Aim:** To study the internal structure of the atom by bombarding a thin gold foil with high-energy, positively charged alpha (alpha) particles.
- **Observation:** * Most alpha particles passed straight through the foil.
    - Some particles were deflected by small angles.
    - A very small fraction (1 in 12,000) rebounded back (180^0).
- **Conclusion:** * **Empty Space:** Most of the atom is empty space.
    - **Concentrated Mass:** All the positive charge and nearly all the mass of the atom are concentrated in an incredibly small volume at the center (the nucleus).
    - **Size Ratio:** The radius of the nucleus is about 10^5 times smaller than the radius of the atom.

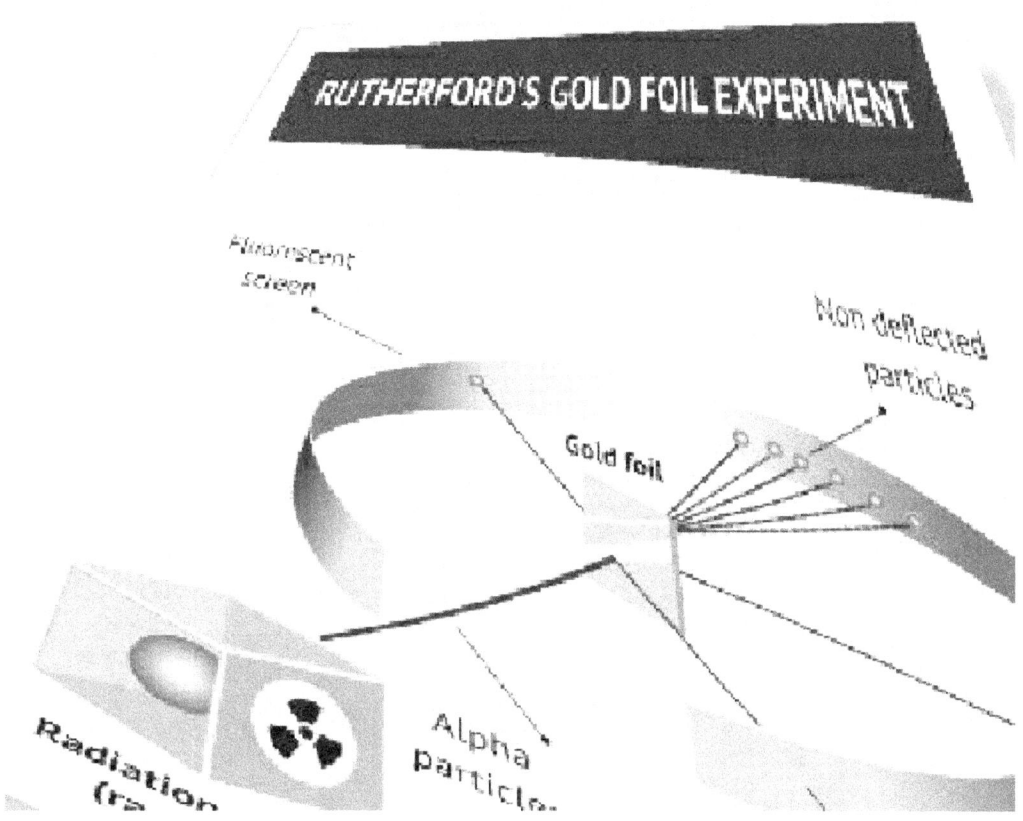

CHAPTER 2
ATOM AND MOLECULES

2.1 LAW OF CHEMICAL EQUATION

Molecules have independent existence. Only these particles actually take part in chemical equation. Atoms have no any independent existence. In every reaction the law of chemical equation is followed. Law of chemical equation consists of two basic laws which are discussed in the following sections:

2.2 LAW OF CONSTANT PROPORTION

In any chemical reaction the constituents' elements are in proportion of their masses. For example: if take water from different source like sea, tape water, formed in the lab, rain and stored in fields, each will have the mass ration of hydrogen to oxygen as 1:8.

2.3 LAW OF CONSERVATION OF MASS

Law of conservation of mass states that in chemical reactions mass can neither be created nor be destroyed. In a given chemical reaction, the mass of reactants as well as mass of product is conserved.

THREE FUNDAMENTAL LAWS OF CHEMICAL REACTIONS: AN ILLUSTRATIVE SUMMARY

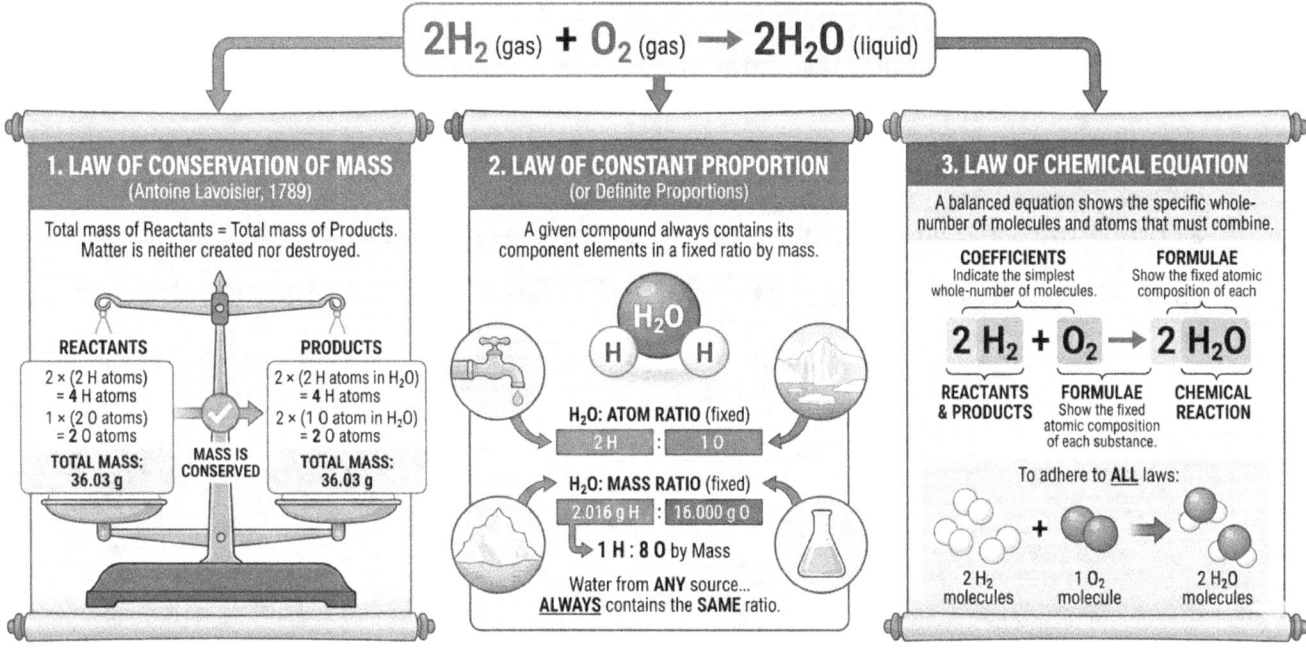

2.4 DALTON'S ATOMIC THEORY

For the first time Dalton gave the term "atom". He suggested that the matter, whether it is element, compound or mixture consists of a small particle called
atom. About atom he gave some postulates generally referred as Dalton's theory of atom. In a simple way we can understand them as follows:

(a) Matter is made up of very small particle known as atom.

(b) We can't create or destroy atom.

(c) Atoms of a particular element are similar in chemical properties as well as atomic masses.

(d) Atoms of different elements have different atomic masses as well as chemical properties.

(e) In a compound the atoms combine in simple ratios of whole number.

(f) In a compound the relative number and kinds of atom is constant.

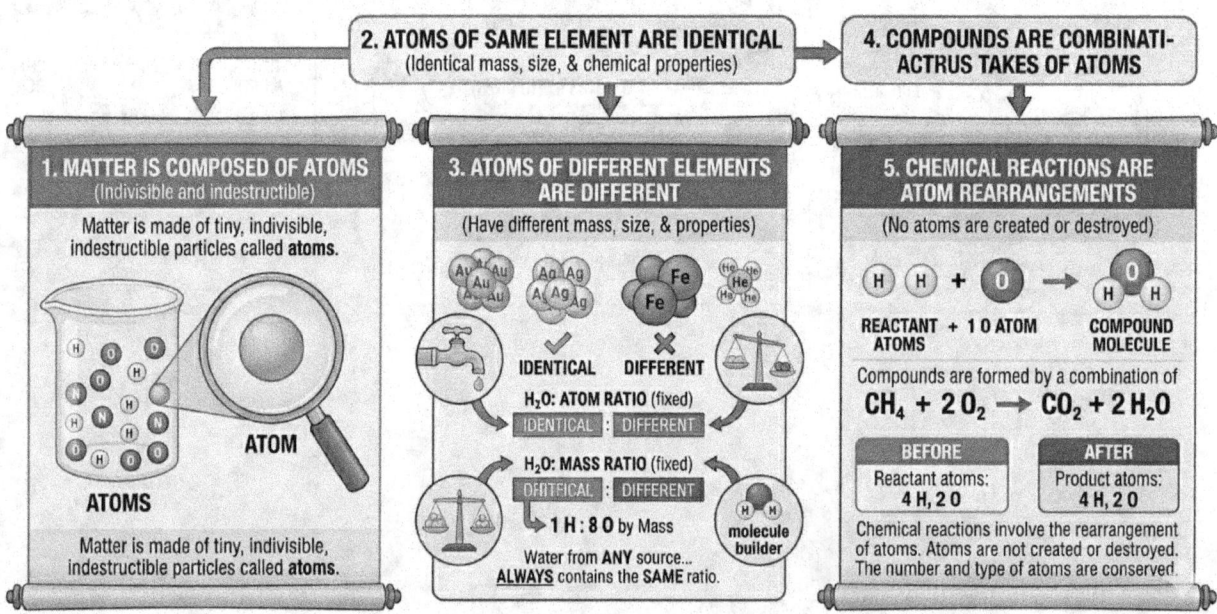

2.5 ATOMIC MASS

For a given atom the atomic mass. refers to the sum of the no. of protons and the neutrons.

2.6 FORMULA MASS AND MOLECULAR MASS

Formula mass refers to the sum of the masses of the total atoms present in the given formula of a compound. Similarly, the molecular mass is the sum of the masses of the atoms present in the given molecule.

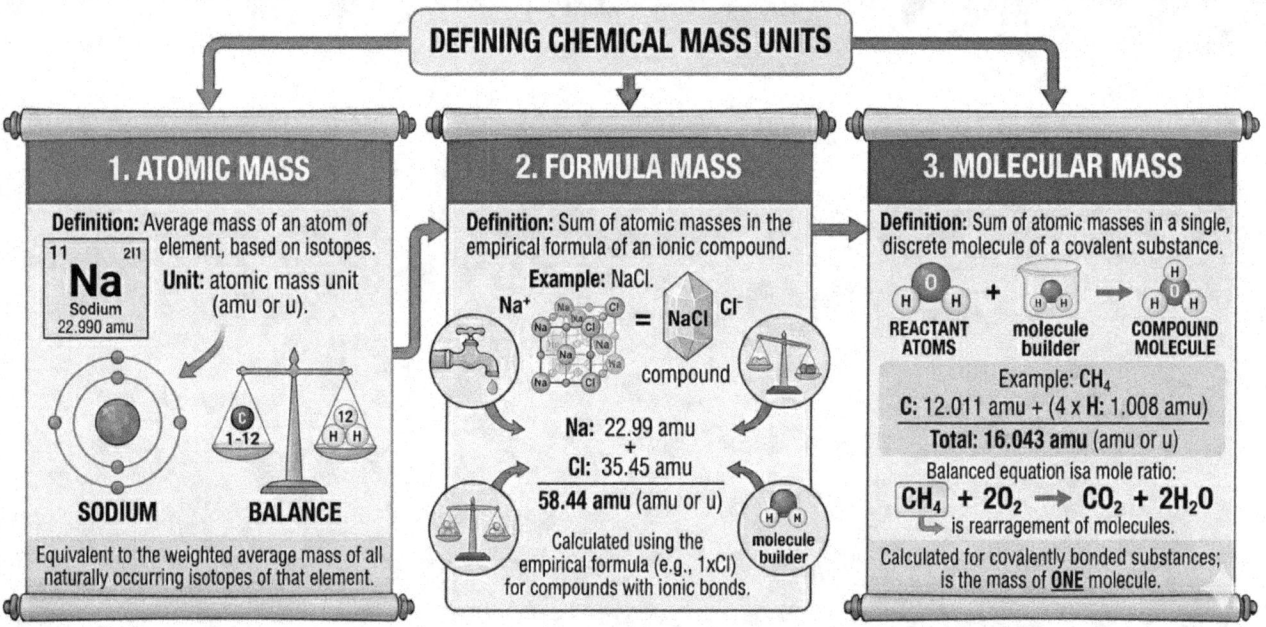

2.7 CONCEPT OF MOLE

In the above sections we have studied that the size of atom and molecules. In real life if we take a half tea spoon of any chemical compound, we can't say how many atoms or molecule are present there? In general life we handle a large no. of atoms and molecule. In this situation the mole concept gives a handy rule to estimate the no. of atoms and molecule in a given quantity of chemical.

In a dozen of banana there are 12 bananas. In a pair of shoes there are 2 shoes. As dozen and pair refers to numbers 12 and 2 respectively. In a similar manner the mole refers to the number 6.022×10^{23}. But now the problem is finding how many moles are there in 4 grams of hydrogen atom. We can understand it in an easy way if we take the no. of gram equal to the atomic mass it will contain one mole. As hydrogen has atomic mass 1 amu so, 1g hydrogen constitutes 1mol of its atom. Similarly, 4g of hydrogen atom will contain 4 moles of hydrogen.

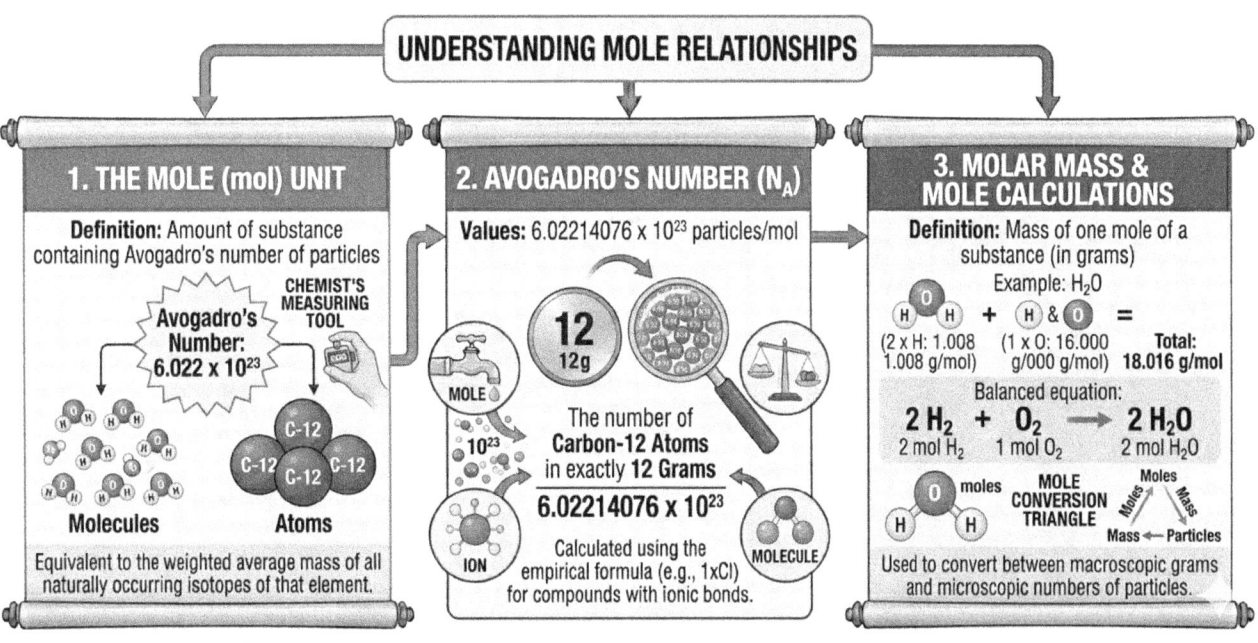

Example 2.3 how many moles of H_2SO_4 molecule is present in 49 grams of Sulphuric acid.

Solution:

Step1:

Here, we have to find the mole of molecule. So, we will first find the molecular mass. The molecular mass of the H_2SO_4 is (2+32+64) 98u.

Step2:

Therefore 98 gram will constitute 1 mol of molecule.

Step3:
Therefore, 49g will constitute (49/98) 0.5 mole.

Molarity

Molarity is used to express the concentration of any given solution.

Molarity = (no. of moles of solute present in solution)/ (volume of solution

CHAPTER 2: DEEP DIVE CHALLENGES

Q1. The word atom was given by?

(a) Bohr

(b) Rutherford

(c) Lavoisier

(d) Dalton

Q2. Law of conservation of mass is given by?

(a) Bohr

(b) Rutherford

(c) Lavoisier

(d) Dalton

Q3. Molecular mass of CuSO4 is given what? Given (Cu=63u, S=32u, O=16u)

(a) 157u

(b) 159g

(c) 159u

(d) 1.59u

Q4. The number of entities in one mole is equal to?

(a) 1.6×10^{-19}

(b) 1.6×10^{-27}

(c) 6.022×10^{23}

(d) None of these

Q5. Total number of electrons present in 44g of CO_2 gas is?

(a) 1 mol

(b) 11 mol

(c) 22mol

(d) None of these

Q6. When 196g of HCl is dissolved in 5-liter waters. What is the molarity of the solution?

(a) 0.2 mol/liter

(b) 1.08 mol/liter

(c) 2 mol/liter

(d) None of these

Q7. How many sodium ions (Na^+) are there in a 58gram of NaCl? (given Na=23u, Cl=35u)

(a) 2 mol/liter

(b) 1 mol/liter

(c) 0.5 mol/liter

(d) none of these

Short Answer Questions

Q1. Briefly explain the following:

(a) Law of conservation of mass

(b) Law of constant proportion

(c) Molar mass

Long answer questions:

Q1. What is Dalton's atomic theory? Explain its postulates.

Solution

1.D 2.C 3.C 4.C 5.C 6. B 7.B

Short Answer

(a) Law of Conservation of Mass
- Mass equality: The total mass of reactants is always equal to the total mass of products in a chemical reaction.
- Rearrangement of atoms: Atoms are neither created nor destroyed; they only change their bonding arrangements.
- Mass of system: In a closed system, the mass remains constant throughout the chemical change.

(b) Law of Constant Proportion
- Fixed mass ratio: Elements are always present in definite proportions by mass in a pure chemical substance.
- Source independence: The composition of a compound remains the same regardless of its origin or method of preparation.
- Chemical purity: A specific compound will always have the same chemical formula and elemental ratio.

(c) Molar Mass
- Gram-atomic/molecular mass: It is the mass of one mole of particles (atoms or molecules) expressed in g/mol.
- Numerical equivalence: The value is numerically equal to the atomic or molecular mass but measured in grams.
- Summation of masses: For a molecule, it is calculated by adding the atomic masses of all its constituent atoms.

Long Answer

Dalton's Atomic Theory Postulates
- Particle Nature: All matter is composed of very tiny, indivisible particles called atoms.
- Indivisibility: Atoms cannot be created, divided, or destroyed in any chemical reaction.
- Elemental Identity: All atoms of a given element are identical in mass and chemical properties.
- Diversity of Elements: Atoms of different elements possess different masses and chemical behaviors.
- Law of Combination: Atoms combine in fixed, simple whole-number ratios to form various compounds.
- Relative Constancy: The relative number and types of atoms are constant in a given compound.

CHAPTER 3
MATTER

3.1 INTRODUCTION

Anything which has mass and occupies space is known as matter. Almost everything surrounding us like air, water, books and food we eat are matter.

3.2 IS AIR MATTER?

Air is essential for our life. We can feel it. But a general quest is that is it matter? In a more formal way has it mass or volume? If we fill a balloon its volume increases so air has volume. A filled balloon has more mass then unfilled so it must have mass. As it has mass as well as volume air is a matter.

3.2 WHAT CONSTITUTES A MATTER?

As atom has no any free existence matter is considered as constitute of molecules.

Properties of matter constituents: the molecule.

The molecules of a matter have some special properties like:

a) They attract each other: In matter the molecules are attached to each other due to inter-molecular forces. In solid it is highest and lowest in gases.

b) There are spaces between them: In any matter there is space between the molecules. These spaces are termed as intermolecular spaces. In solids it is less and highest in gases.
It can be seen in an easy demonstration. When we make a solution of sugar and water, the sugar adjusts in the intermolecular spaces of the water and sugar is not seen in the solution.

c) They are continuously moving: whether it is solid, liquid or gases the molecules are in continuous motion.

For example, due to this property we are able to smell of a tasty food in other room.

3.3 STATES OF MATTER

3.4 SOLID

In solid the intermolecular force is maximum and the intermolecular spaces least. For this reason, solids have fixed shape and volume.

3.5 LIQUID

In liquid the intermolecular force is less than solids but greater than liquids. The intermolecular spaces are more than solid but less than gases. For this reason
liquid has fixed volume. This intermolecular force is not enough to fix the shape and size. Therefore, the liquid takes the shape of the container in which it is kept.

3.6 GASES

In the gases the intermolecular spaces are most and the intermolecular force is least. That's why molecules are free to move in entire space. Gases have no fixed volume nor shape and size.

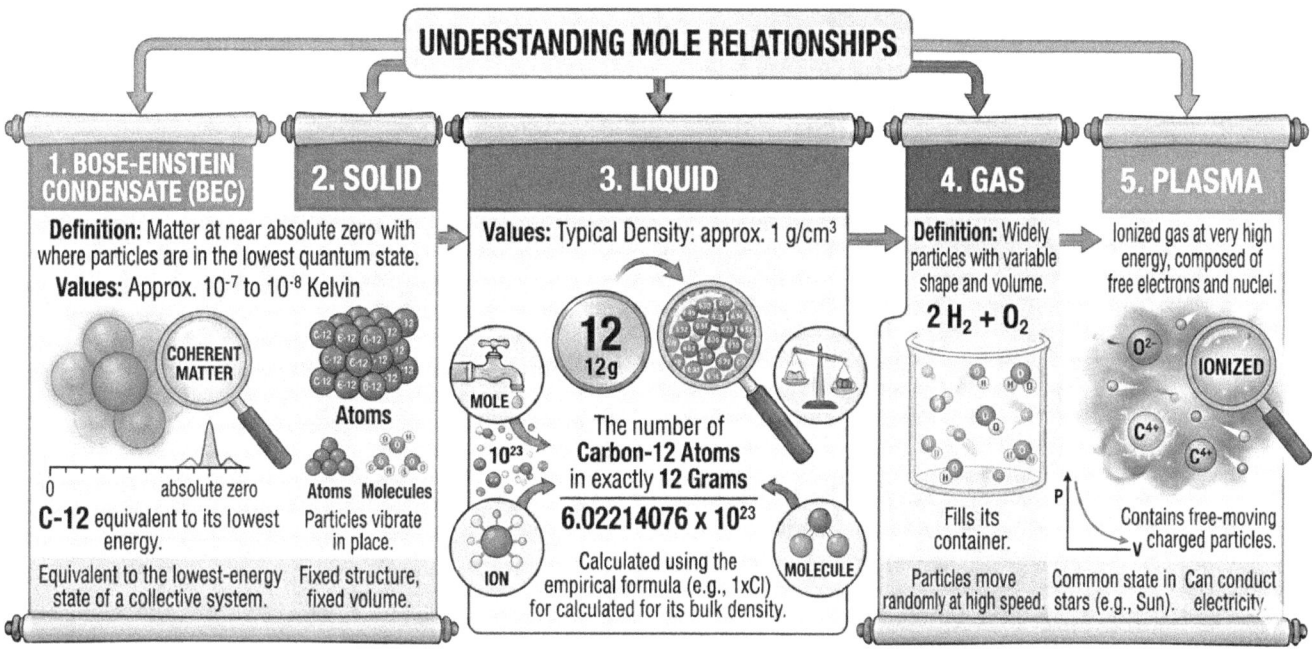

3.7 PLASMA

Plasma is the fourth state of matter in which the negative and positive ions are mixed and form the matter. These are loosely bonded in comparison to the gases. Approx charge on the plasma is zero.

3.8 BOSE EINSTEIN CONDENSATE (BEC)

The fifth state is Bose Einstein condensate. Its compactness is much more than a solid state. Generally, it is the gases cooled at the 0 Kelvin.

3.9 CHANGING OF STATES OF MATTER

Solid, liquid and gases are transformed from one state to other due to change in temperature and pressure.

Melting: When heat is added to the solid it changes to the liquid. This process is known as melting.

Freezing: Freezing is the reverse process of the melting. In this process Liquid changes to the solid.

Condensing: In condensing process gases are changed to the solid on decrease of temperature.

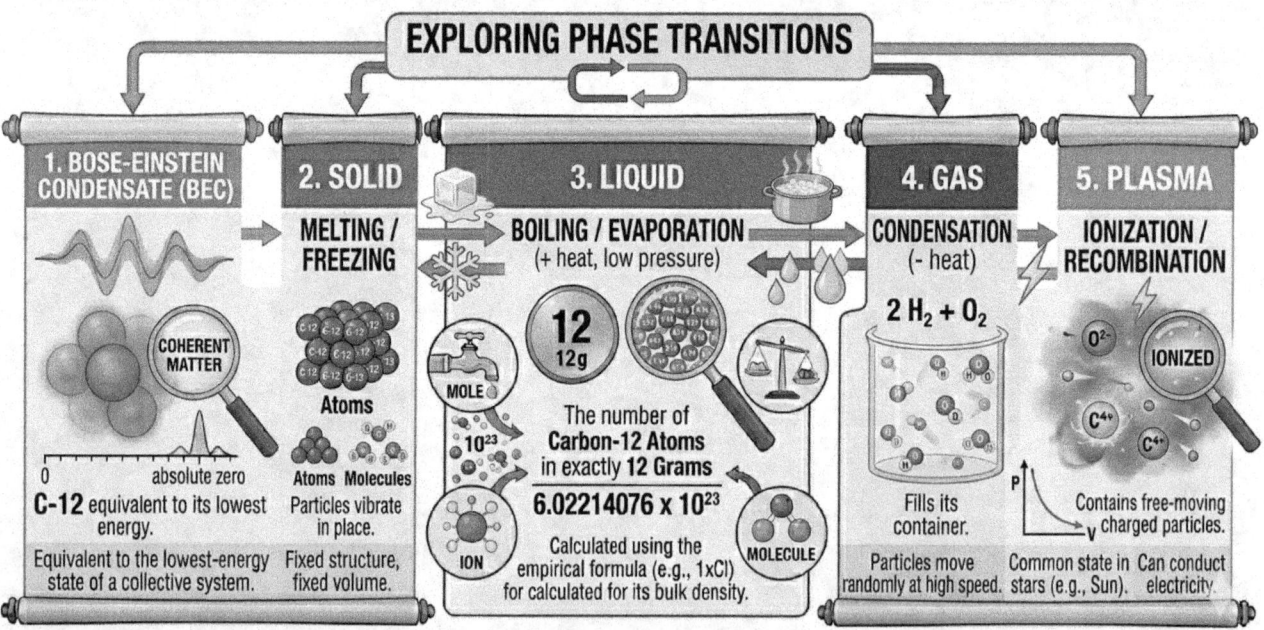

Boiling: In boiling the liquid is converted into gaseous state on the account of heat gain.

Sublimation: Sublimation is the single process which is responsible for the inter conversion of the solid to gases and vice-versa.

Ionization: Gaseous state is converted to the plasma state when the gaseous atoms are ionized.

Deionization: Deionization is responsible for the conversion of plasma to the gaseous state.

3.10 EVAPORATION: A PROCESS OF COOLING

Definition: below the boiling point of any liquid, the process of converting liquid into vapor is known as evaporation.

In an open liquid surface, there are many particles or molecule which have enough energy to break the intermolecular forces and changes into vapor. This process is may be at any temperature and generally referred as the evaporation. It is a cooling process because in evaporation the lost energy of the liquid surface is gained from surrounding and the surrounding cool

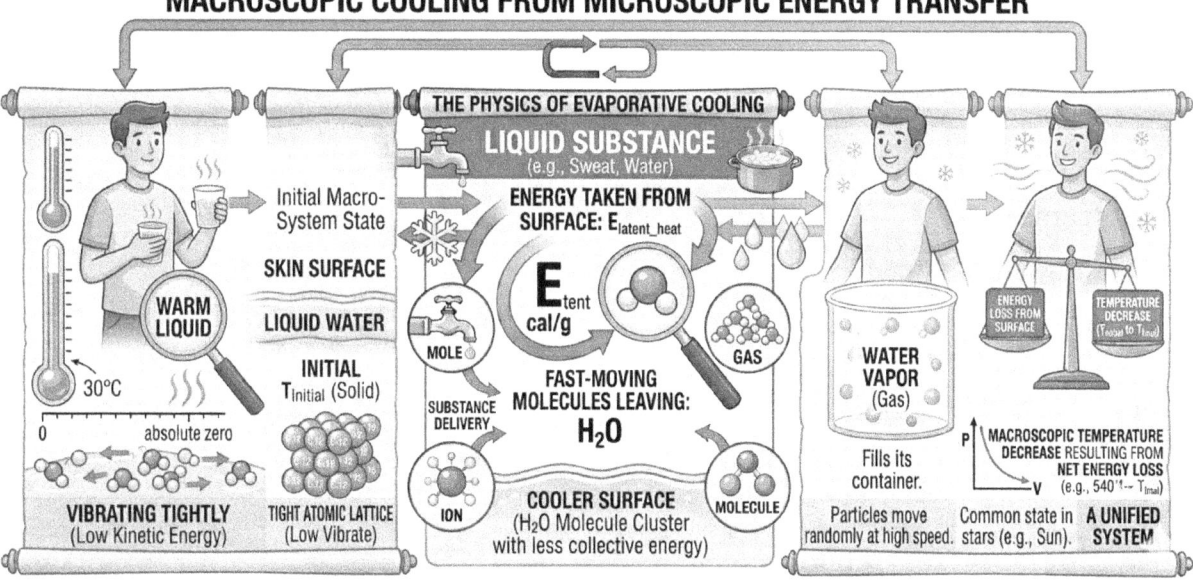

CHAPTER 3: DEEP DIVE CHALLENGES

Q1. Which has fixed shape and volume?

(a) Solid

(b) Liquid

(c) Gas

(d) Plasma

Q2. In which intermolecular space is Maximum?

(a) Solid

(b) Liquid

(c) Gas

(d) all has same

Q3. Diffusion is involved in?

(a) Solid

(b) Liquid

(c) Gas

(d) None of these

Q4. The process of changing solid to gas directly is known as?

(a) Evaporation

(b) Condensation

(c) Sublimation

(d) Melting

Q5. On adding heat, the intermolecular forces between two molecules:

(a) Increases

(b) Decreases

(c) First increases then decreases

(d) None of these

Short Answer Questions:

Q1. Compare the properties of solid, liquid and gases.

Q2. What is the difference between melting and sublimation?

Q3. How can you demonstrate that "oxygen is matter"?

Solution

---
1. A   2. C   3. C   4. C   5. B

Q1. Comparison of Solid, Liquid, and Gas
- Shape and Volume: Solids have a fixed shape and volume; liquids have a fixed volume but no fixed shape; gases have neither a fixed shape nor a fixed volume.
- Compressibility: Solids are incompressible, liquids have very low compressibility, and gases are highly compressible due to large intermolecular spaces.
- Fluidity: Solids are rigid and do not flow; liquids and gases are fluids and can flow easily.

Q2. Difference between Melting and Sublimation
- Melting: The process in which a substance changes from a solid state to a liquid state upon heating (e.g., ice turning into water).
- Sublimation: The process in which a substance changes directly from a solid state to a gaseous state (or vice versa) without passing through the liquid phase (e.g., camphor or dry ice).

Q3. Demonstration that "Oxygen is Matter"
- Definition of Matter: Matter is anything that has mass and occupies space.
- Mass: By weighing an empty oxygen cylinder and then weighing it after it is filled with oxygen, the increase in weight proves that oxygen has mass.
- Volume: When oxygen is blown into a balloon, the balloon expands. This inflation demonstrates that the gas occupies space (volume) inside the balloon.

CHAPTER 4
PURE AND IMPURE MATTER

4.1 INTRODUCTION

Matters surrounding us can be separated in two groups i.e. pure and impure. Pure matters have fixed composition while impure have no fixed composition. Further pure substances can be divided into two group's elements and compounds.

Elements cannot be further broken into simpler substances. Examples of elements are iron, copper, sodium, uranium etc. however compounds can be broken into simpler substances by chemical and electrochemical processes.

Impure matter has no fixed composition however may have uniformity in distribution such mixture is termed as homogeneous mixture. Example of homogeneous is water in alcohol. In a general form we can say that the elements and compounds are pure matter and mixtures are impure.

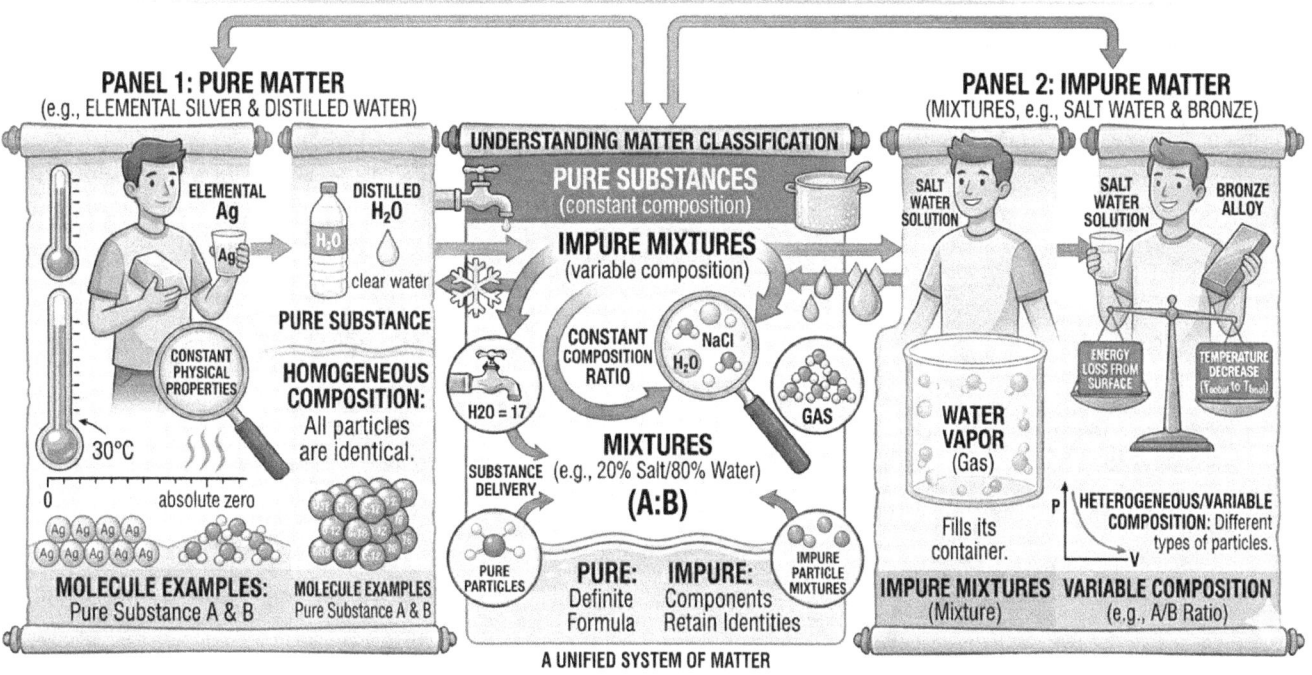

4.2 PHYSICAL AND CHEMICAL CHANGES

Physical change only involves the change of states. In chemical change the property of the matter also changes and in addition heat and light evolves.
Changing of ice into liquid is physical change while rusting of iron is chemical change.

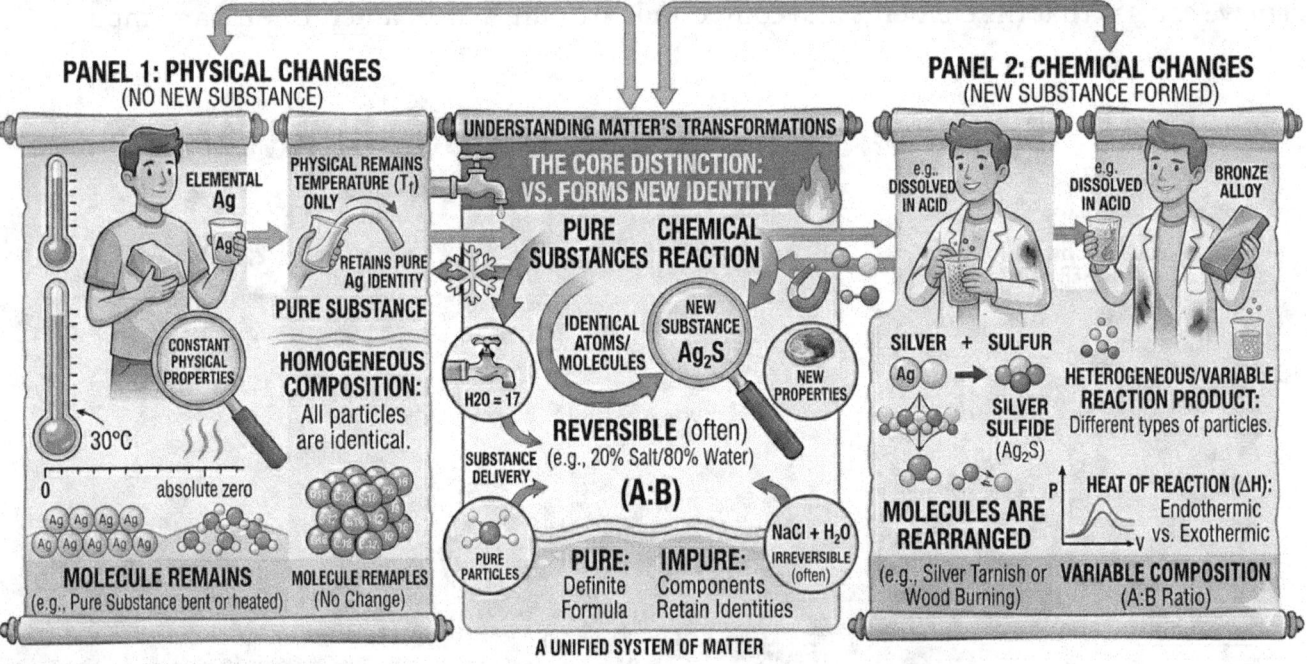

4.3 MIXTURE

When two or more than two elements or compounds combine physically in any proportion mixture is formed.

Mixture can be homogeneous or heterogeneous. If the constituents of mixture are uniformly distributed over the entire range the mixture is called homogeneous mixture. Sugar solution is the example of homogeneous mixture. In heterogeneous mixture the distribution of constituent is non uniform. Mixture of sand and rice is the example of heterogeneous mixture.

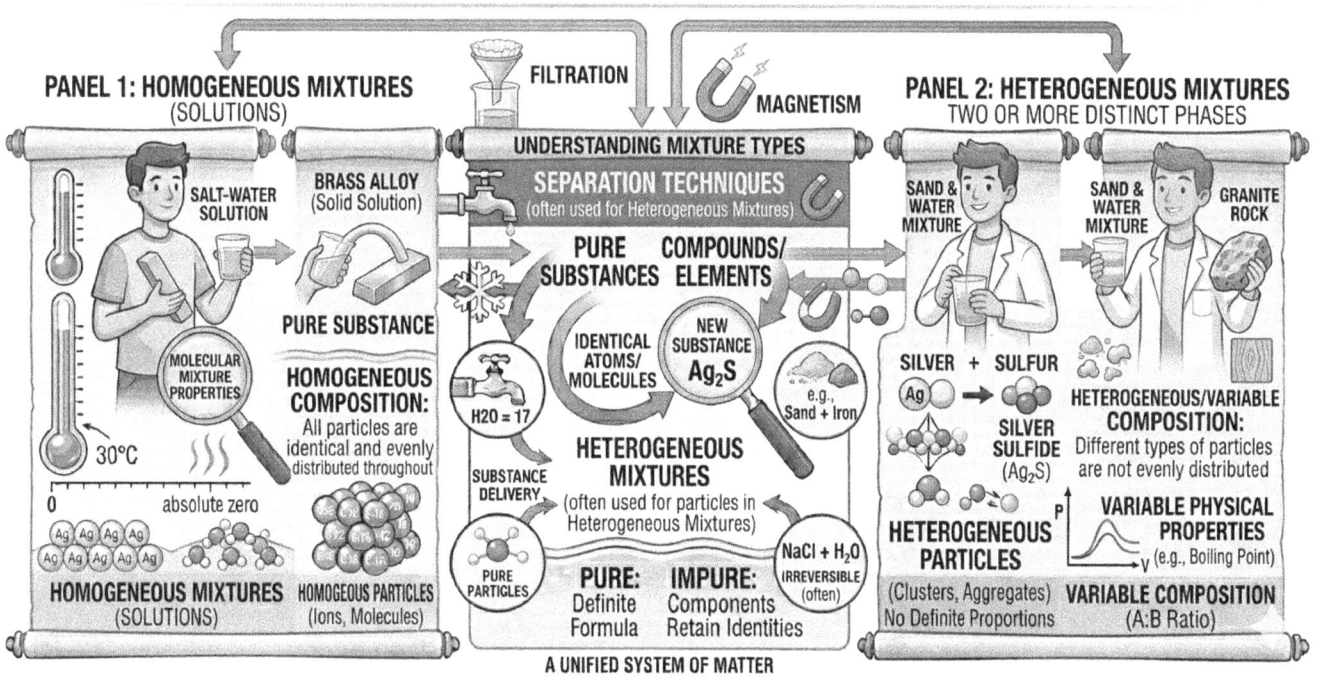

4.4 Solution

Solution is a homogeneous mixture. It may be the mixture of two solids like alloy, two or more gases like air. It may also be the homogeneity of more than two states like sugar dissolved in water. The constituent which dissolves is known as solvent and which is dissolved called solute. Solutes can't be filtered out.

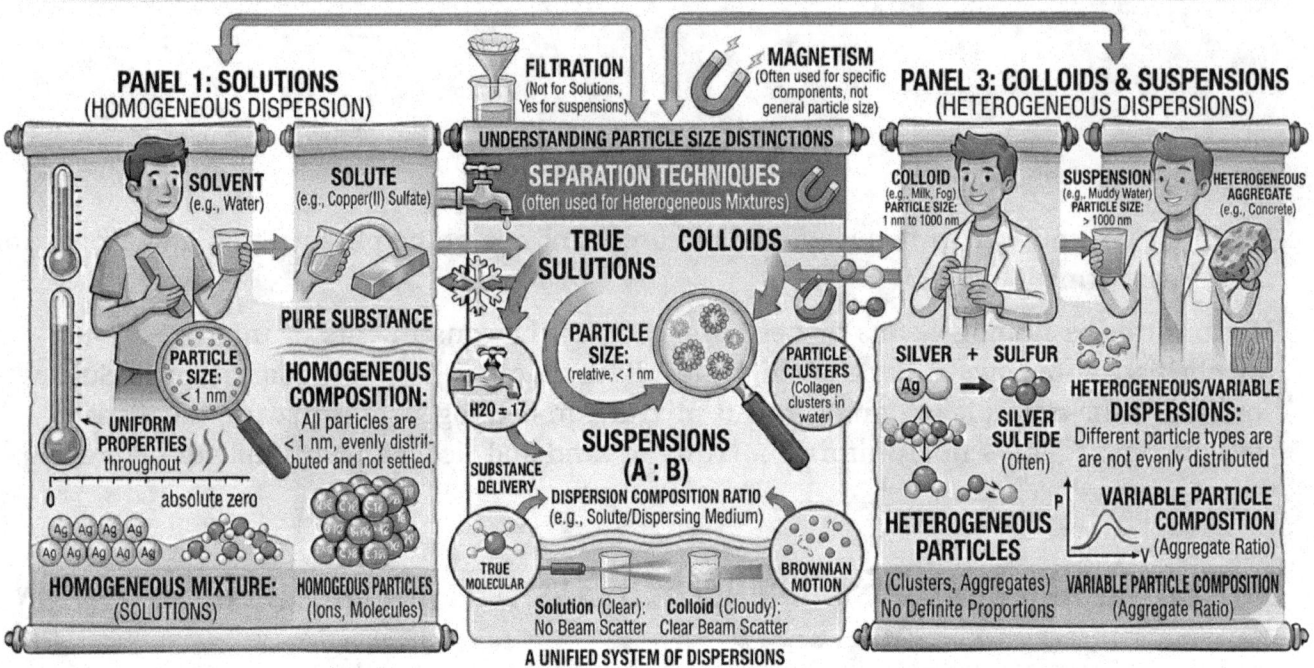

4.5 PROPERTIES OF SOLUTION

1. In a solution the solutes are of the order less than nano metre. So, we can't see them by our naked eye.

2. It can't scatter the light passing through them.

3. Solutes cannot settle down when kept the solution in undisturbed state.

4.6 EXPRESSING CONCENTRATION OF SOLUTION

(i) Mass percentage of a solution = (Mass of solute/Mass of solution) x 100

(ii) Volume percentage of a solution = (Mass of solute/Volume of solution) x 100

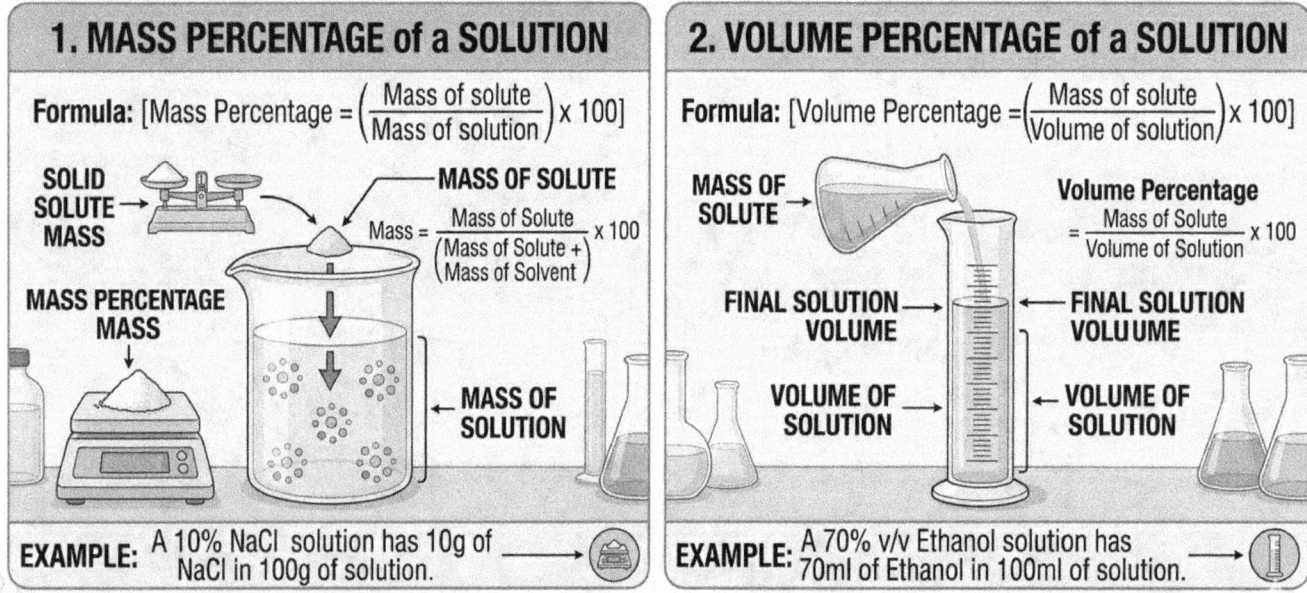

4.7 SUSPENSION

A suspension is heterogeneous mixture. Its solute can be seen by naked eyes. It scatters the light and shows a visible path. it is not stable like solution because in undisturbed conditions its solutes settle down and the suspension behavior is suspended. Its solute can be separated by using filter.

4.8 COLLOIDAL

In colloidal the size of solute is in between the solution and the suspension. A special name dispersed phase for solute and dispersing medium for solvent is used. The dispersed phases are such that they can't be filtered out. However
dispersed phase can be separated using centrifugation technique. They also show the scattering of light. Colloids are named as per their phase and medium. Some examples are illustrated below

4.9 EXAMPLE OF COLLOIDS

CLASSIFICATION OF TYPES OF COLLOIDS

DISPERSED PHASE	DISPERSING MEDIUM	TYPE OF COLLOID	EXAMPLE & ILLUSTRATION
LIQUID	GAS	AEROSOL	Fog, Clouds, Mist
SOLID	GAS	AEROSOL	Smoke, Automobile Exhaust
GAS	LIQUID	FOAM	Shaving Cream
LIQUID	LIQUID	EMULSION	Milk, Face Cream
SOLID	LIQUID	SOL	Milk of Magnesia, Mud
GAS	SOLID	FOAM	Foam Rubber, Sponge, Pumice
LIQUID	SOLID	GEL	Jelly, Cheese, Butter
SOLID	SOLID	SOLID SOL	Colored Gemstone, Milky Glass

COLLOIDS: VITAL HETEROGENEOUS MIXTURES IN DAILY LIFE

4.10 COMPOUND

When two or more than two elements combine chemically in a definite proportion it results into a compound. For example, $2H_2$ molecules combine with single oxygen molecule O_2 to form two molecules of H_2O.

4.11 DIFFERENCES BETWEEN MIXTURE AND COMPOUNDS

Mixture

1. The resultant mixture is not a new compound.

2. It has none uniform composition.

3. The resultant mixture has the properties of the constituent element or compounds.

4. The components can be separated using simple physical methods.

Compound

1. The resultant is a new compound.

2. It has uniform and fixed composition.

3. The resultant compound has the different properties from its constituting elements.

4. Its components can be separated only using chemical or electrochemical processes.

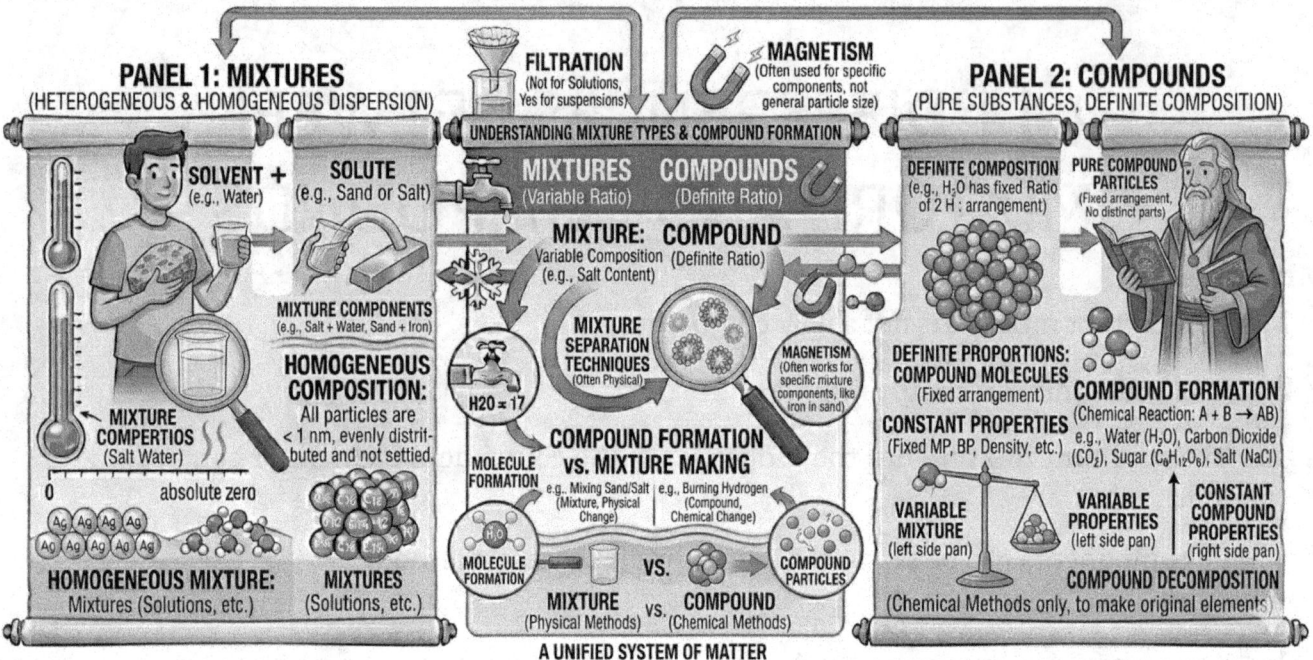

4.12 SOME METHODS TO SEPARATE THE COMPONENTS FROM MIXTURE: EVAPORATION

It is used to separate the mixture of volatile and non-volatile substances.

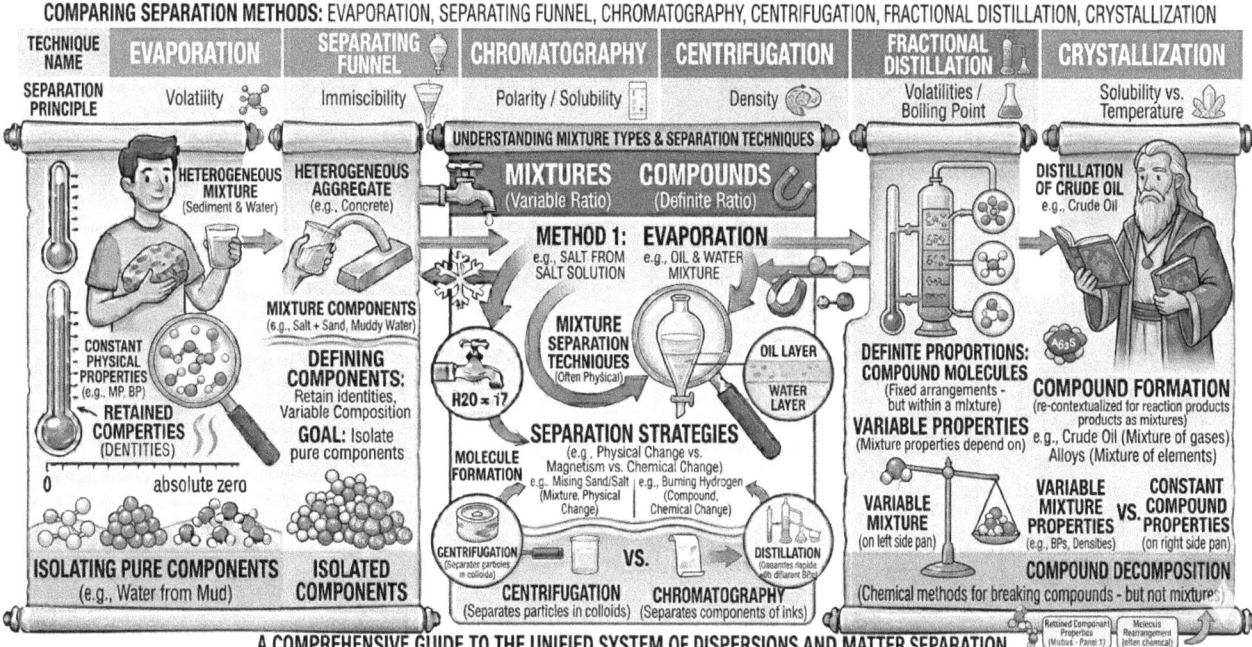

4.13 CHROMATOGRAPHY

It is used to separate the solutes which are dissolved in the same solvent. Applications

To separate
a) Colors in a dye
b) Pigments from natural colors
c) Drugs from blood

4.14 SEPARATING FUNNEL

It is used to separate two immiscible liquids like water and oil.

4.15 CENTRIFUGATION

It is based on the principle that when a mixture of solid in liquid is spun rapidly the lighter one come on the top and denser to bottom.

Applications

a) In diagnostic laboratories for blood and urine tests.
b) In dairies and home to separate butter from cream.
c) In washing machines to squeeze out water from wet clothes.

4.16 FRACTIONAL DISTILLATION

It is used to separate the mixture of two miscible liquids. Petroleum products are separated from this technique. It is also useful in separating different gases from the air. For this purpose, we decrease temperature and increase pressure to form a liquid air. Liquid air is then fractionally distilled to get the different gases.

4.17 SUBLIMATION

It is used to separate solids from mixture which sublime i.e. ammonium chloride, camphor, naphthalene and anthracene.

4.18 CRYSTALLIZATION

It is the process of purifying solids.

Applications

a) Purification of sea salt.
b) Separation of crystals of alum from impure samples.

CHAPTER 4: DEEP DIVE CHALLENGES

Q1. NaCl is a?

(a) Pure substance

(b) Impure substance

(c) Mixture of pure and impure

(d) None of these

Q2. Solution of sugar in water is?

(a) Element

(b) Compound

(c) Homogeneous mixture

(d) Heterogeneous mixture

Q3. Milk is a?
(a) Colloidal

(b) Solution

(c) Suspension

(d) Compound

Q4. Cream is obtained from milk using which processes.

(a) Filtration

(b) Chromatography

(c) Sublimation

(d) Centrifugation

Q5. Which process is used to separate drugs from blood?

(a) Filtration

(b) Chromatography

(c) Sublimation

(d) Centrifugation

Q6. Which process is used to squeezes water from cloth in washing machine?

(a) Filtration

(b) Chromatography

(c) Sublimation

(d) Centrifugation

Solution

1. A 2. C 3. A 4. D 5. B 6. D

NOTES

PART 2

CHAPTER 5

CHEMICAL REACTIONS AND EQUATIONS

5.1 INTRODUCTION

In our day-to-day life we come across many reactions. Rusting of iron articles, formation of curd from milk and digestion of food in our body all are the examples of chemical reactions. For every reaction there are some common features that are observed. In a chemical reaction one or more following properties must be seen. These are:

a. Change in state
b. Change in color
c. Evolution of gases
d. Change in temperature

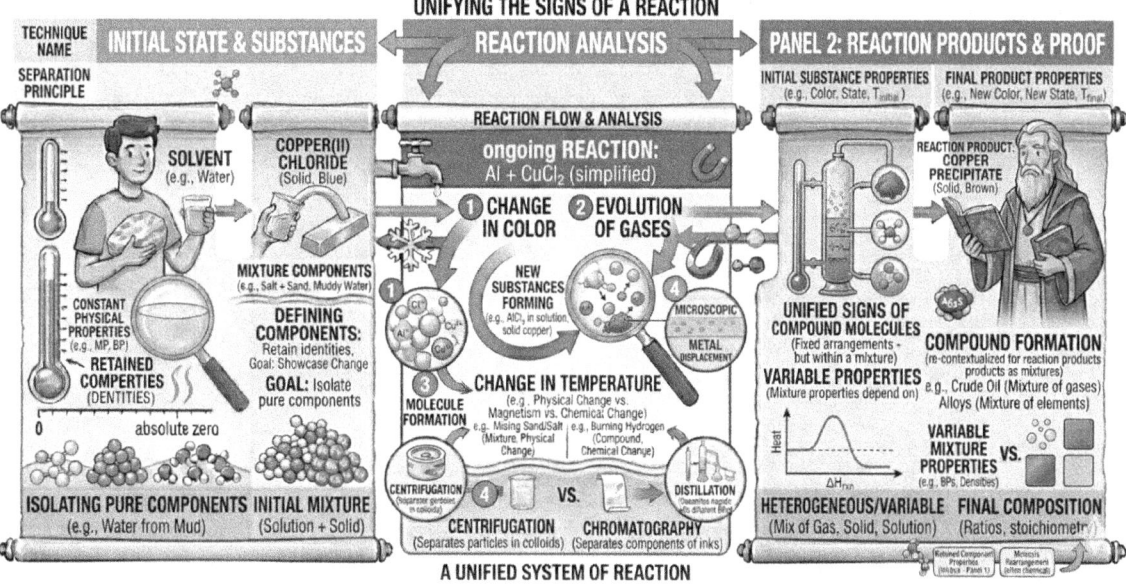

Now, it is easily observable that during the rusting of iron like rime of our bicycle its color changes to brown. it is the change of color. During the formation of
curd liquid changes to solid. In the digestion of food almost all changes like change of state, change of color, change of temperature and evolution of gases take place.

5.2 CHEMICAL REACTION AND EQUATION

Chemical reaction is an actual process while chemical equation is the symbolic representation of chemical reaction.

In a chemical equation the substances which take part are known as reactants and the substances which are formed referred as the product. In an ideal chemical equation, the reactants are kept left side of arrow and added to each other by plus (+) sign and products are kept right side of the arrow.

In a chemical equation the state of the reactants and products are also shown i.e. Solid (s), liquid (l), gas (g), aqueous (aq), precipitate (ppt).

5.3 BALANCED CHEMICAL EQUATION

In a balanced chemical equation, the number of atoms of each element is equal in reactant sides as well as product side.

If a given reaction is not balanced it is called *skeletal equation*.

5.4 BALANCING A CHEMICAL EQUATION

Method 1: giving priority to oxygen and hydrogen

For balancing a chemical equation if we give priority to oxygen and then hydrogen it becomes quite easier to balance.

Example 3: balance the following reaction $Fe + O_2 \rightarrow Fe_2O_3$

Here we observe that in LHS there are two atoms of oxygen and 3 atoms of oxygen in RHS. If we multiply LHS by 3 and RHS by 2 the oxygen becomes balanced.

$Fe + 3O_2 \rightarrow 2Fe_2O_3$

Now,

We see that oxygen is balanced then for Fe it is 1 in LHS and 4 in RHS therefore multiply by 4 in LHS and reaction becomes balanced.

$4Fe + 3O_2 \rightarrow 2Fe_2O_3$

Method 2: an algebraic method

If you are quite interested in mathematics. Here is a handy method for you. Example 4: balance the following equation
$Fe + O_2 \rightarrow Fe_2O_3$

Step1:

Let the coefficient of Fe, O_2 and Fe_2O_3 be a, b and c respectively. Then equation becomes

$a\, Fe + b\, O_2 \rightarrow c\, Fe_2O_3$

Step2:

let it is balanced equation then a=2c
and 2b=3c
here c has the maximum multipliers so let c=1. On solving it we get
A=2 B=1.5
C=1
Then equation becomes $2Fe + 1.5 O_2 \rightarrow Fe_2O_3$

Step3:

For getting integral coefficients we can multiply by 2 in LHS and RHS. We get
4Fe+3O$_2$ → 2Fe$_2$O$_3$

Note: in many cases step 3 doesn't require.

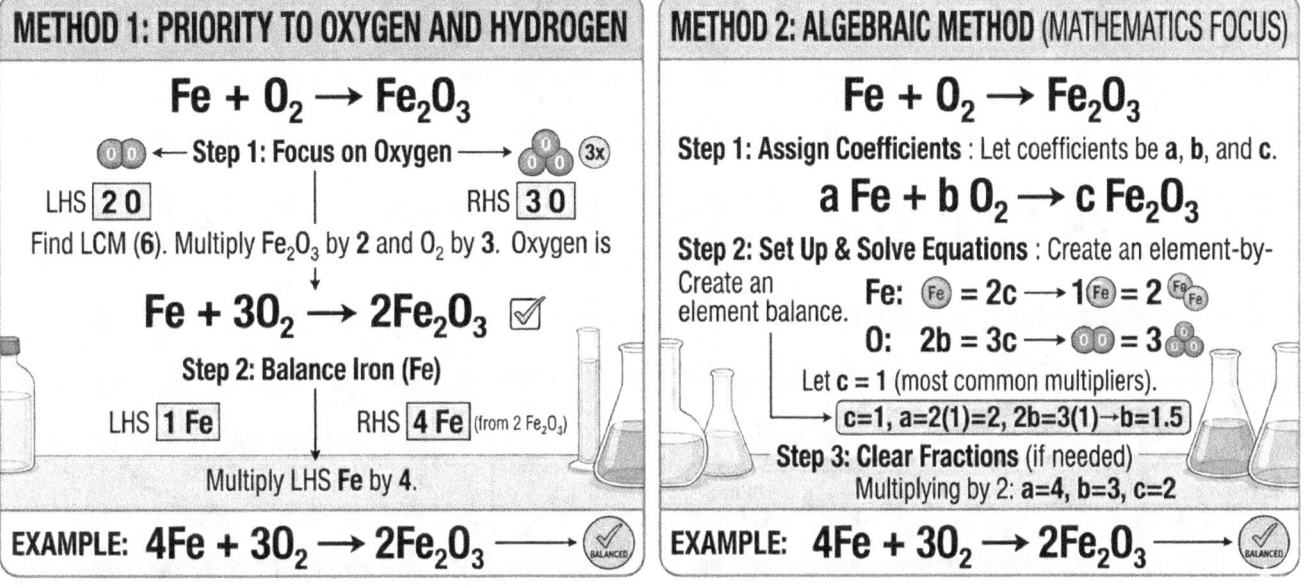

5.5 TYPES OF CHEMICAL REACTIONS

1. Combination reaction: when two or more than two reactants combine to form a single product, it is known as combination reaction.

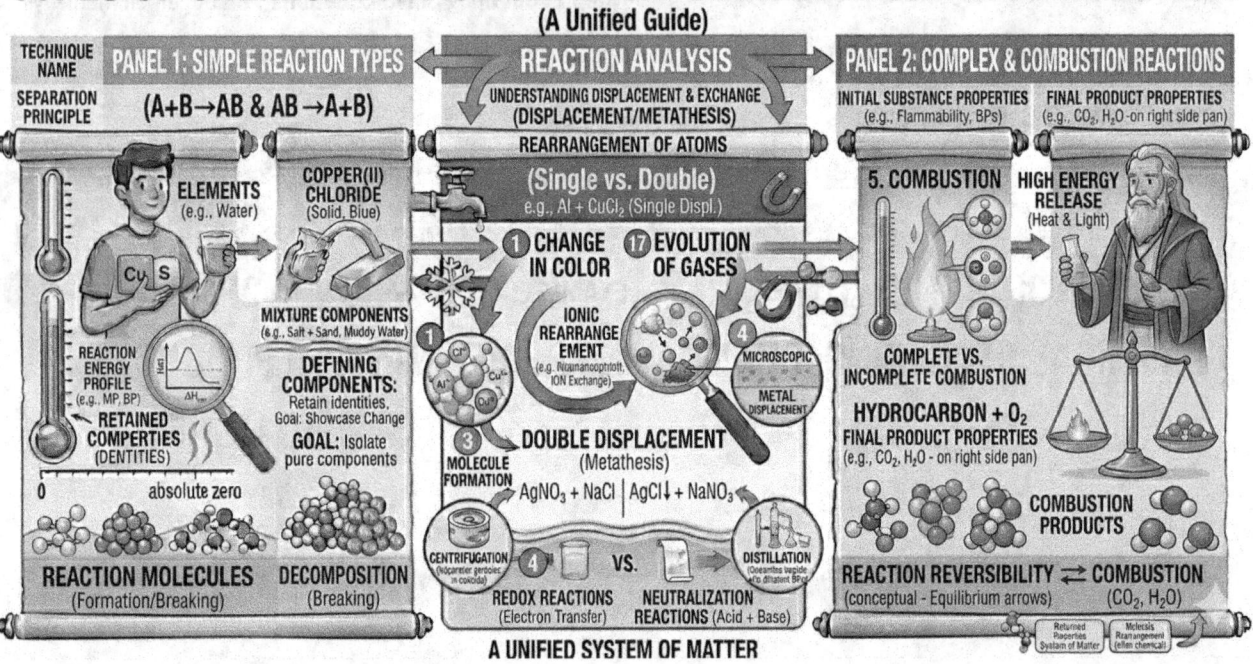

2. Exothermic reaction: the reaction in which heat is released along with the formation of product is known n as exothermic reaction.

Respiration and digestion of food are best example for this.

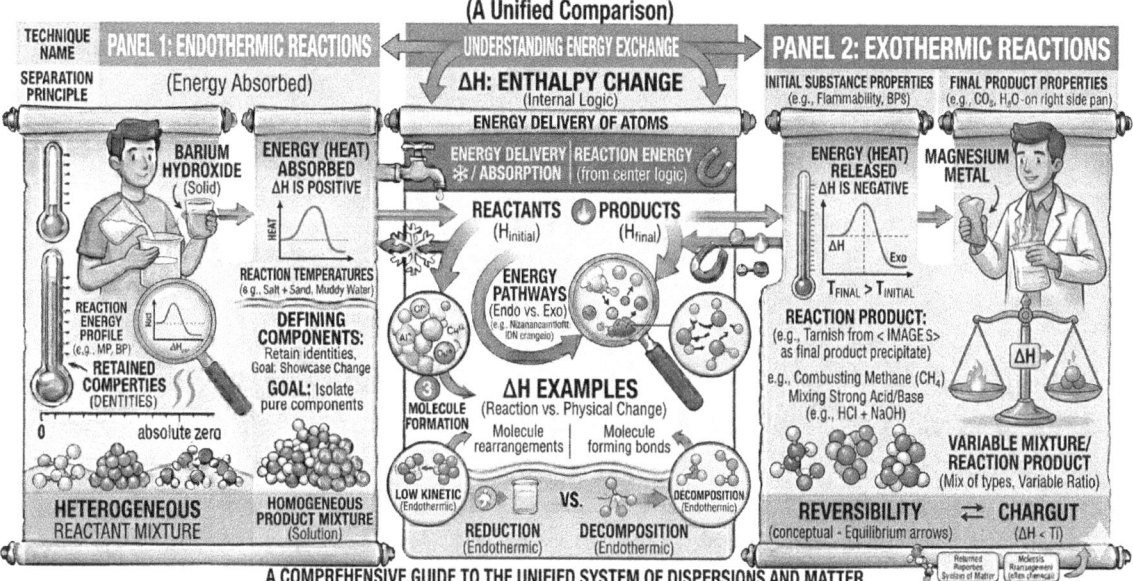

3. **Endothermic reaction:** Endothermic reaction is the reverse process of exothermic. In this reaction the heat is absorbed along the formation of the products.

4. **Decomposition reaction:** in decomposition reaction a single reactant breaks into two or more than two products. This reaction can be carried out using heat or electricity and respectively referred as thermal or electrical decomposition reaction.

5. **Displacement reaction:** this reaction is genuinely reffers to the exchange of ions in when an ionic compound reacts with another higher reactive metal. In actual practice the higher reactive element displaces the lower from its ion.

6. **Double Displacement reaction:** it is similar to the displacement reaction only difference is that in this reaction both reactants are ionic compounds. In this reaction the ions are exchanged.

7. **Precipitation reaction:** in this type of reaction the precipetate is formed which is setteled down to the bottom. This reaction can be also identified in the balanced chemical equation by noticing a downward arrow.

$$Na_2SO_4 \text{ (aq)} + BaCl_2 \text{ (aq)} \rightarrow BaSO_4 \text{ (ppt)} + 2NaCl \text{ (aq)}$$

8. **Oxidation and Reduction reaction:** in a general way oxidation refers to the addition of oxygen or loss of hydrogen and reduction refers to the loss of oxygen or gain of hydrogen. Actually, addition of oxygen removes electron and loss of oxygen refers to the gaining of electron.

In a simple way we can state that in a particular reaction if any atom gaining electron, it is reduced or if it is removing electron it is oxidizing. Oxidation and reduction are the supplementary to each other. Both of these processes are taking place in single reaction, because if there is removal of electrons certainly removed electron is added to another atom participating in the reaction.

5.6 REACTIONS AFFECTING OUR LIFE

Corrosion

When any metal comes into the contact of air, water, moisture and acid, it corrodes and this process is known as the corrosion. It is very harmful for our countries economy. In a single year due to corrosion many things made up of iron and other metals corrode and government has to repair it. For example, the government buses, monument's safety bar made up of iron. Our door, the rim of motor vehicles these are also affected.

Corrosion can be minimized using oil, grease on the surface of metallic part, it prevents the air, water, moist and acid to come into the contact of the metal surface. Galvanization is another process to prevent it. In this process a layer of zinc is added to the surface of other metals like iron. As zinc is less prone to corrosion, it is a suitable choice. This done using electrolysis processes.

Rancidity

When oily substances like chips comes into the contact of oxygen get oxidized and their tastes changes. This process is referred as the rancidity. Avoiding these chips packets are flushed with nitrogen gases.

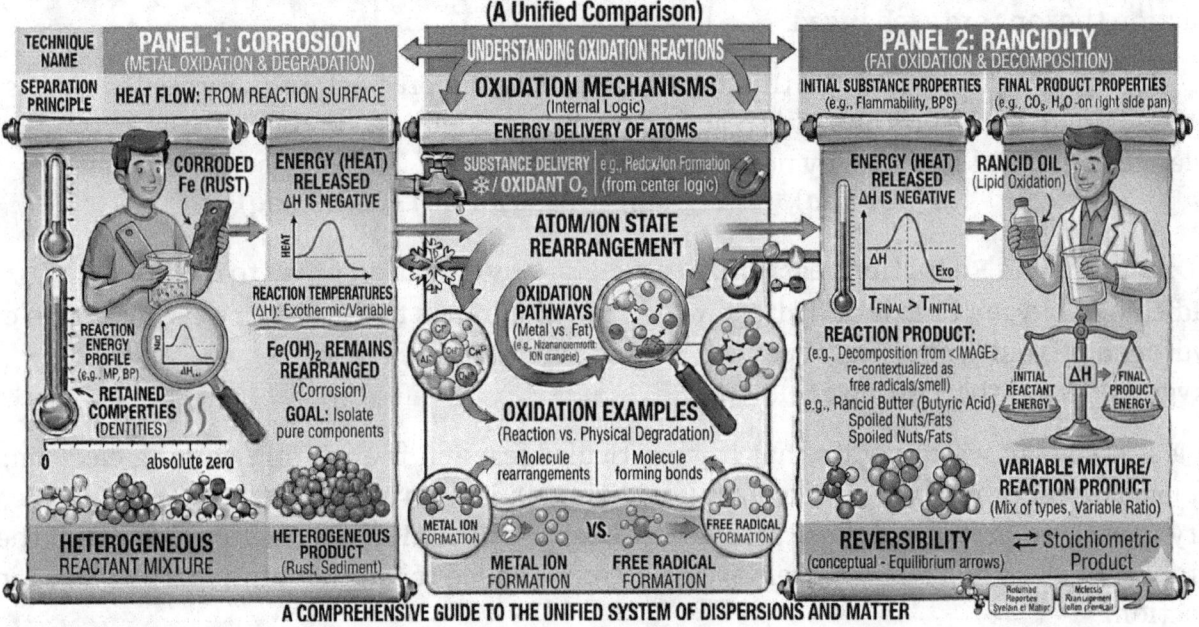

CHAPTER 5: DEEP DIVE CHALLENGES

Q1. When Zinc reacts with HCl which gas is released?

(a) H_2
(b) He
(c) O
(d) Cl

Q2. In equation the given reaction, which is reactant?

$H_2 + O_2 \rightarrow H_2O$

(a) H_2
(b) O_2
(c) Both
(d) None

Q3. $Fe + H_2O \rightarrow ? + H_2$

Which compound should be at the place of ? in the above equation?

(a) Fe_2O_3
(b) FeO
(c) Fe_3O_4
(d) $Fe(OH)_3$

Q.4 $3Fe + 4H_2O \rightarrow ? + 4H_2$

Which compound should be at the place of? in the above equation?

(a) Fe_2O_3
(b) FeO
(c) Fe_3O_4
(d) $Fe(OH)_3$

Q5. $C + O_2 \rightarrow CO_2$ is?

(a) Displacement reaction

(b) Endothermic reaction

(c) Exothermic reaction

(d) None

Q6. What is the formula of lime stone?

(a) CaO

(b) CaCO2

(c) Ca(OH)2

(d) CaCO3

Q7. what is the formula of quick lime?

(a) CaO

(b) CaCo2

(c) Ca(OH)2

(d) CaCO3

Q.8 what should be the value of X so that equation becomes balanced?
2Pb(NO3)2(s) → 2PbO(s) + XNO2(g) + O2(g)

(a) 1

(b) 2

(c) 3

(d) 4

Q.9   Fe(s) + CuSO4(aq) → FeSO4(aq) + Cu(s)
Which type of reaction is this?

(a) Displacement reaction

(b) combination reaction

(c) decomposition reaction

(d) None

Q.10   which is oxidized in the given reaction?

2Cu + O2 → CuO

(a) Cu

(b) O2

(c) CuO

(d) None

Q.11 Which is essential for corrosion?

(a) air

(b) water

(c) moisture

(d) all of these

Q.12 silver articles become black when exposed to air. In this process which compound is formed?

(a) Silver oxide

(b) Silver nitrate

(c) Silver sulphide

(d) Silver carbonate

Q13. Food kept in air tight jar prevent it from?

(a) oxidation

(b) reduction

(c) corrosion

(d) None

Q14. Chips packets are flushed with?

(a) N_2

(b) O_2

(c) Cl_2

(d) He

Q.15 $2Cu + O_2 \rightarrow 2CuO$
Which type of reaction is this?

(a) Displacement reaction

(b) combination reaction

(c) decomposition reaction

(d) None

Q16. In which type of chemical reaction ions are interchanged?

(a) Displacement reaction

(b) combination reaction

(c) decomposition reaction

(d) double displacement reaction

Q17. Opposite of combination reaction is?

(a) Displacement reaction

(b) combination reaction

(c) decomposition reaction
(d) double displacement reaction

Q18. Oxidation is the gain of?
   (a) oxygen
   (b) electron
   (c) proton
   (d) none of these

Q19. Reduction is the gain of?
   (a) oxygen
   (b) electron
   (c) proton
   (d) hydrogen

Q20. Precipitates are?
   (a) Soluble in water
   (b) Insoluble in water
   (c) Partially soluble
   (d) Float in water

Match the followings:

Q21.

A	B
$AgNO_3$	Combination reaction
$2H_2 + O_2 \rightarrow 2H_2O$	Black
$Fe_2O_3 + Al \rightarrow Al_2O_3 + Fe$	unbalanced reaction
rancidity	nitrogen

Q22.

A	B
oxidation	gain of electron
reduction	loss of electron
corrosion	chips
rancidity	iron

Q23.

A	B
Black and white photography	thermite
reaction Exothermic reaction	AgBr
$CuSO_4$	Au
Noble metal	blue

Q24. What is a balanced chemical equation?

Q.25 what is the difference between combination reaction and decomposition reaction?

Q26. what is the difference between exothermic reaction and endothermic reaction?

Q27. what is the difference between displacement reaction and double displacement reaction?

Q28. What are essential conditions for corrosion?

Q29. Why chips packets are flushed with nitrogen gases?

Q30. Write down the reaction which is generally used in black and white photography?

Q31. What is rancidity? Explain with examples.

Q32. Why we apply paints on metal articles?

Q33. A shiny brown colored element 'X' on heating in air becomes black in color. Name the element 'X' and the black-colored compound formed.

Long answer type questions

Q34. Translate the following statements into chemical equations and then balance them.

(a) Hydrogen gas combines with nitrogen to form ammonia.

(b) Hydrogen sulphide gas burns in air to give water and Sulphur dioxide.

(c) Barium chloride reacts with aluminum sulphate to give aluminum chloride and a precipitate of barium sulphate.

(d) Potassium metal reacts with water to give potassium hydroxide and hydrogen gas.

Q35. Balance the following chemical equations.

(a) $HNO_3 + Ca(OH)_2 \rightarrow Ca(NO_3)_2 + H_2O$

(b) $NaOH + H_2SO_4 \rightarrow Na_2SO_4 + H_2O$

(c) $NaCl + AgNO_3 \rightarrow AgCl + NaNO_3$

(d) $BaCl_2 + H_2SO_4 \rightarrow BaSO_4 + HCl$

Q36. Write the balanced chemical equations for the following reactions.

(a) Calcium hydroxide + Carbon dioxide → Calcium carbonate + Water

(b) Zinc + Silver nitrate → Zinc nitrate + Silver

(c) Aluminum + Copper chloride → Aluminum chloride + Copper

(d) Barium chloride + Potassium sulphate → Barium sulphate + Potassium chloride

Q37. Write the balanced chemical equation for the following and identify the type of reaction in each case.

(a) Potassium bromide(aq) + Barium iodide(aq) → Potassium iodide(aq) + Barium bromide(s)
(b) Zinc carbonate(s) → Zinc oxide(s) + Carbon dioxide(g)
(c) Hydrogen(g) + Chlorine(g) → Hydrogen chloride(g)
(d) Magnesium(s) + Hydrochloric acid(aq) → Magnesium chloride(aq) + Hydrogen(g)

Solution

1.A 2.C 3.A 4.C 5.C 6. A 7.C 8. B 9.A 10. A 11.D 12. B 13.A 14. A 15.B 16. D 17.C 18. A

19.B 20. B

Short Answer Hints
- Q24: Think about the Law of Conservation of Mass. Are the number of atoms for each element the same on the left and right?
- Q25: One is about "joining" ($A + B \rightarrow AB$), while the other is about "breaking down" ($AB \rightarrow A + B$).
- Q26: Focus on the flow of heat. Does the test tube feel hot (releasing energy) or cold (absorbing energy)?
- Q27: In one, a "stronger" element kicks out a "weaker" one. In the other, two compounds simply exchange partners.
- Q28: Metals need two "partners" from the environment to rust: one is a liquid and one is a gas found in the air.
- Q29: Oxygen causes food to spoil. Nitrogen is inert (unreactive) and acts as a protective cushion.
- Q30: Look for the decomposition of silver halides (Silver Chloride or Silver Bromide) when exposed to sunlight.
- Q31: This involves the oxidation of fats and oils in food, leading to a bad smell and taste. Think of old, "stale" butter.
- Q32: Paint acts as a barrier. If air and moisture can't touch the metal, what process is prevented?
- Q33: Element 'X' is a common metal used in electrical wires. When it reacts with oxygen (O_2), it forms a black oxide.

Long Answer Hints
Q34: Translation and Balancing
- a) $H_2 + N_2 \rightarrow NH_3$. (Hint: You need 2 Nitrogen atoms on the right).
- (b) $H_2S + O_2 \rightarrow H_2O + SO_2$. (Hint: Oxygen appears in two places on the right; count

BASIC CHEMISTRY

carefully!)
- (c) $BaCl_2 + Al_2(SO_4)_3 \rightarrow AlCl_3 + BaSO_4$. (Hint: Balance the Sulfate (SO_4) groups as a single unit).
- (d) $K + H_2O \rightarrow KOH + H_2$. (Hint: Use a coefficient of 2 for K, H_2O, and KOH).

Q35: Balancing Equations
- (a) Look at the Nitrate (NO_3) group; there are two on the right.
- (b) Start by balancing the Sodium (Na) atoms first.
- (c) Check the atoms carefully—is this one already balanced?
- (d) Balance the Chlorine (Cl) and Hydrogen (H) by adding a single coefficient on the right.

Q36: Writing Balanced Equations
- (a) $Ca(OH)_2 + CO_2 \rightarrow CaCO_3 + H_2O$.
- (b) Zinc is more reactive than Silver; it will take the Nitrate group.
- (c) Aluminum has a valency of 3, while Copper is usually 2. This affects the subscripts in $AlCl_3$ and $CuCl_2$.
- (d) This is a partner-exchange reaction. Ensure the formulas for Barium Sulfate and Potassium Chloride are correct.

Q37: Balancing and Identification
- (a) Type: Double Displacement. (Hint: Ions are being exchanged).
- (b) Type: Decomposition. (Hint: A single reactant is breaking into two products).
- (c) Type: Combination/Synthesis. (Hint: Two elements are forming one compound).
- (d) Type: Displacement. (Hint: A metal is reacting with an acid to release a gas).

CHAPTER 6
ACIDS, BASES AND SALTS

6.1 FORMATION OF ACIDS

When nonmetals react with oxygen it produces nonmetallic oxide.

$S(s) + O_2(g) \rightarrow SO_2(g)$

$P_4(s) + 5O_2(g) \rightarrow P_4O_{10}(s)$

$2H_2(g) + O_2(g) \rightarrow 2H_2O(l)$

When nonmetallic oxide reacts with water acid is formed.

$SO_3(g) + H_2O(l) \rightarrow H_2SO_4(aq)$

$P_4O_{10}(s) + 6H_2O(l) \rightarrow 4H_3PO_4(aq)$

$4NO_2(g) + 2H_2O(l) + O_2(g) \rightarrow 4HNO_3(aq)$

6.2 FORMATION OF BASES

When metals react with oxygen it produces metallic oxide.

$$4Fe + 3O_2 \rightarrow 2Fe_2O_3$$
$$2Cu + O_2 \rightarrow 2CuO$$
$$4Na + O_2 \rightarrow 2Na_2O$$
$$4Al + 3O_2 \rightarrow 2Al_2O_3$$

When metallics oxide reacts with water base is formed.

$CaO(s) + H_2O(l) \rightarrow Ca(OH)_2(aq)$
$K_2O(s) + H_2O(l) \rightarrow 2KOH(aq)$
$MgO(s) + H_2O(l) \rightarrow Mg(OH)_2(aq)$

In a general way we can understand the formation of base by a very simple activity.

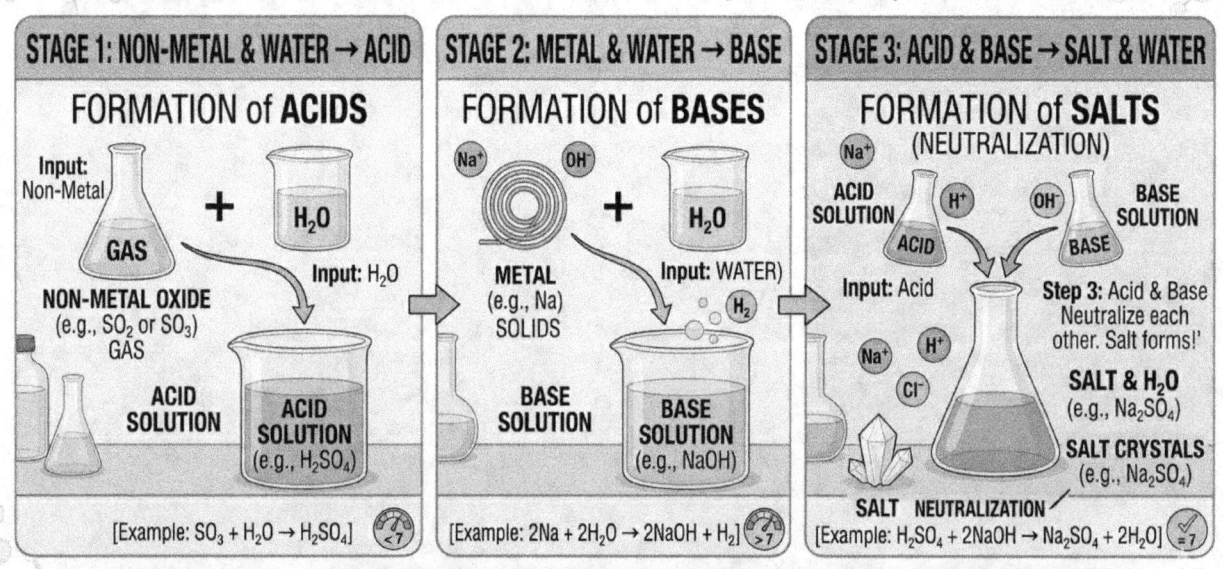

Activity 1:

Step 1: take a magnesium ribbon and clean its upper surface by sand paper.
Step2: burn it with the help of a pair of tongs and a spirit lamp.
Step 3: collect its ash and dissolve in a glass of water.
Now you have a base in your glass.

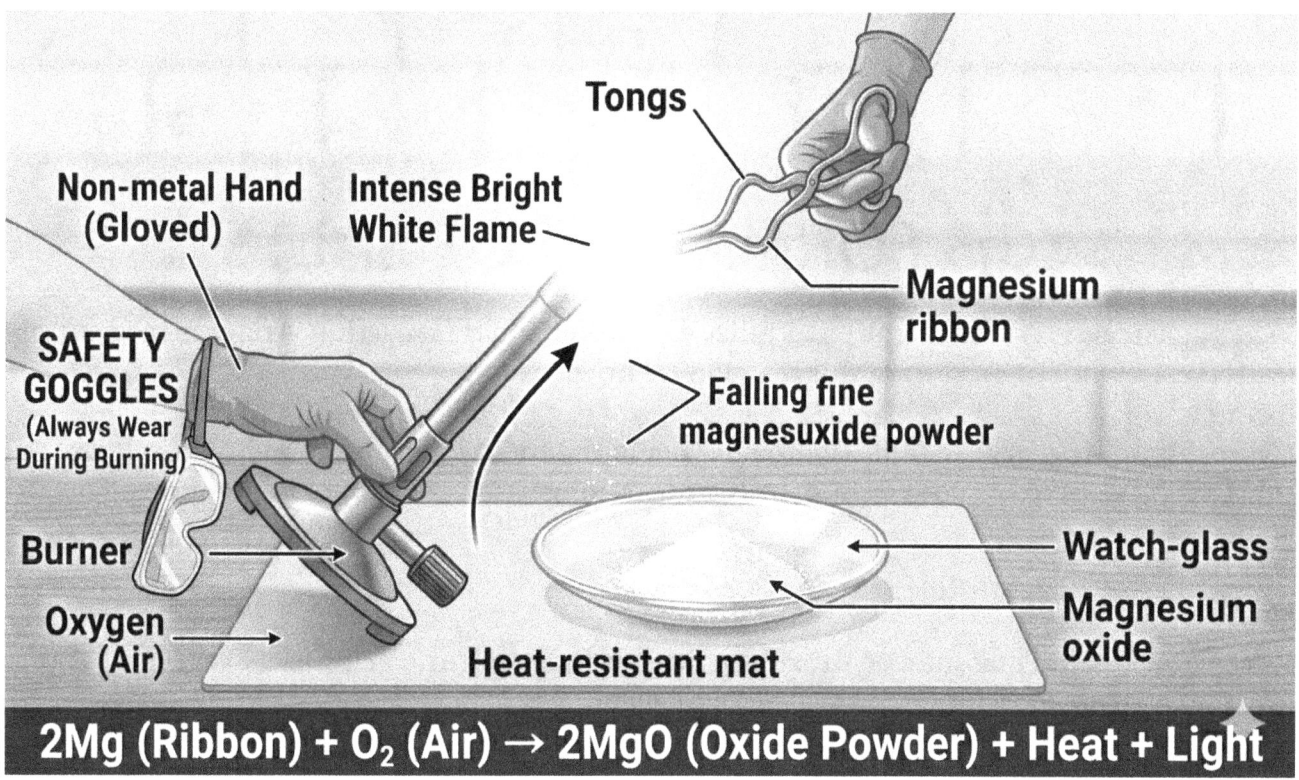

6.3 FORMATION OF SALTS

When acid reacts with base, both neutralize each other and this reaction is referred as neutralization reaction. The product of this reaction is known as salt.

Example:

Hydrochloric Acid and Sodium Hydroxide This is the most classic example. When hydrochloric acid (strong acid) reacts with sodium hydroxide (strong base), they produce common table salt and water.

Reaction: $HCl(aq) + NaOH(aq) \rightarrow NaCl(aq) + H_2O(l)$

Sulfuric Acid and Potassium Hydroxide In this reaction, sulfuric acid reacts with potassium hydroxide to form potassium sulfate (a salt) and water.

Reaction: $H_2SO_4(aq) + 2KOH(aq) \rightarrow K_2SO_4(aq) + 2H_2O(l)$

Nitric Acid and Calcium Hydroxide When nitric acid reacts with calcium hydroxide (slaked lime), it produces calcium nitrate and water.

Reaction: $2HNO_3(aq) + Ca(OH)_2(aq) \rightarrow Ca(NO_3)_2(aq) + 2H_2O(l)$

Now we have a general idea of formation of acids, bases and salt. In the next section we will discuss some physical and chemical properties of acid, base and salt.

6.4 DETECTION OF ACIDS AND BASES IN GENERAL LIFE

Generally, acids are sour in taste while bases are bitter in nature. Acid turns blue litmus paper to red while base changes red to blue. When acids and bases are in solution or aqueous form we can check whether they are acids or bases using many natural flowers and petals. These flowers or petals are referred as natural indicator. Some of these are red cabbage, turmeric, colored petals of flowers such as hydrangea, petunia and geranium. There are many synthesized indicators. These are shown in the following table.

DETECTION OF ACIDS & BASES IN GENERAL LIFE

Sour Taste → Acid (e.g., Lemon) →[H⁺] Turns Blue Litmus to Red

NATURAL INDICATORS: Can also be detected using natural flowers & petals (e.g., red cabbage, turmeric, hydrangea, petunia, geranium)

Changes Red Litmus →[OH⁻] Base (e.g., Soap) → Bitter to Blue

Indicator	pH range (color change)	color of acid form	color of conjugate base form
methyl orange	3,1 - 4,4	orange	yellow
methyl red	4,2 - 6,2	red	yellow
bromothymol blue	6,0 - 7,8	yellow	blue
phenolphthalein	8,3 - 10,0	colourless	pink
alizarin yellow	10,1 - 12,1	yellow	red

6.5 CHEMICAL PROPERTIES OF ACIDS AND BASES

Reaction with metals

When acids or bases react with metal salt and hydrogen gas is produced.

$$NaOH + Zn \rightarrow Na_2ZNO_2 + H_2$$

Reaction with metal carbonates and hydrogen carbonates

On reaction with metal carbonate or metal hydrogen carbonate acids and bases produces salt and water with carbon di oxide gas.

1. Reaction with Metal Carbonates.

   The general equation is:
   Acid + Metal Carbonate → Salt + Water + Carbon Dioxide

   Example: Hydrochloric Acid and Sodium Carbonate When dilute hydrochloric acid is added to sodium carbonate, sodium chloride, water, and carbon dioxide are formed. Reaction: $Na_2CO_3(s) + 2HCl(aq) \rightarrow 2NaCl(aq) + H_2O(l) + CO_2(g)$

2. Reaction with Metal Hydrogen Carbonates The general equation is:
   Acid + Metal Hydrogen Carbonate → Salt + Water + Carbon Dioxide

   Example: Hydrochloric Acid and Sodium Hydrogen Carbonate When hydrochloric acid reacts with sodium hydrogen carbonate (baking soda), the products are the same as above. Reaction: $NaHCO_3(s) + HCl(aq) \rightarrow NaCl(aq) + H_2O(l) + CO_2(g)$

How to Test for Carbon Dioxide (CO_2)

To confirm that the gas produced is carbon dioxide, it is passed through lime water (calcium hydroxide). The lime water turns milky due to the formation of a white precipitate of calcium carbonate.

Test Reaction: $Ca(OH)_2(aq) + CO_2(g) \rightarrow CaCO_3(s)\downarrow + H_2O(l)$

Metallic oxides with acids

As metallic oxide is basic in nature when it reacts with any acid salt is formed with the release of hydrogen gas.

Nonmetallic oxides with base

As nonmetallic oxide is acidic in nature when it reacts with any base salt is formed with the release of hydrogen gas.

Acids and bases in aqueous state

When acid or bases are diluted in water, they release H^+ or H^- ions respectively. These ions are necessary to give authentic pH test or any indicator test. That's why all test relevant to acids and bases are taken in aqueous state only.

The following table will recapitulate the properties of acids and bases:

RECAPITULATING THE PROPERTIES OF ACIDS AND BASES: KEY CHEMICAL REACTIONS

REACTION TYPE	REACTANTS	PRODUCTS	SUBSTANCES FORMULA
Acid-Metal Reaction (Forms Hydrogen Gas)	Acid + Metal	Salt + H_2	e.g., $HCl + Mg \rightarrow MgCl_2 + H_2$
Neutralization (Hydroxide)	Acid + Metal Hydroxide	Salt + H_2O	e.g., $HCl + NaOH \rightarrow NaCl + H_2O$
Neutralization (Oxide)	Acid + Metal Oxide	Salt + H_2O	e.g., $H_2SO_4 + CuO \rightarrow CuSO_4 + H_2O$
Acid-Carbonate Reaction (Forms CO_2)	Acid + Metal Carbonate	Salt + H_2O + CO_2	e.g., $HCl + CaCO_3 \rightarrow CaCl_2 + H_2O + CO_2$
Acid-Bicarbonate Reaction (Forms CO_2)	Acid + Metal Hydrogen Carbonate	Salt + H_2O + CO_2	e.g., $HCl + NaHCO_3 \rightarrow NaCl + H_2O + CO_2$
Neutralization (Acidic Oxide)	Acidic Oxide + Base	Salt + H_2O	e.g., $CO_2 + Ca(OH)_2 \rightarrow CaCO_3 + H_2O$

CONSOLIDATED REACTION SUMMARY

6.6 PH SCALES A STRONG TOOL

pH scale is a scale designed to test the acidic or basic character of any substance. In this scale there are 0 to 14 readings. In this reading 7 refers to the neutral substance while less is acidic and more is basic in nature.

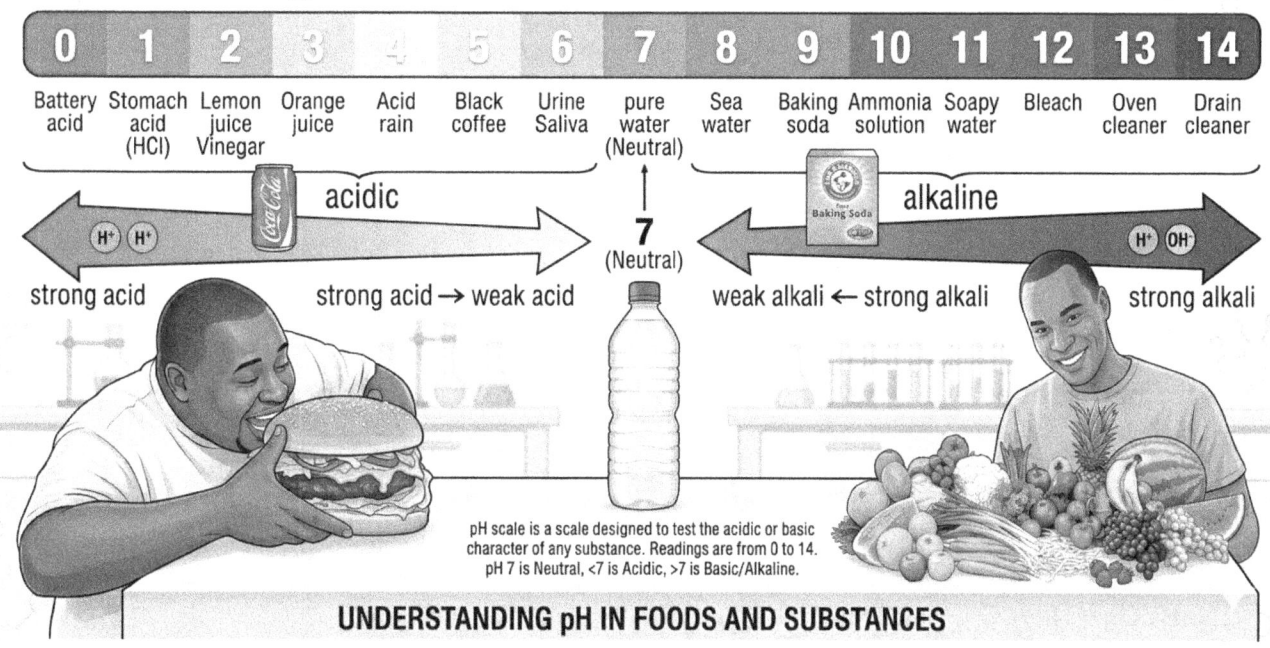

6.7 SOME IMPORTANT SALTS

Bleaching powder

Formation

Chlorine gas is used in the formation of bleaching powder. Bleaching powder is produced by the action of chlorine on dry slaked lime [Ca(OH)2].

$$Ca(OH)_2 + Cl_2 \rightarrow CaOCl_2 + H_2O$$

Uses

(i) for bleaching cotton and linen in the textile industry
(ii) for bleaching wood pulp in paper factories
(iii) for bleaching washed clothes in laundry
(iv) As an oxidizing agent in many chemical industries
(v) For disinfecting drinking water to make it free of germs

Baking soda Formation

In the formation of baking soda i.e. sodium hydrogen carbonates the basic raw material is sodium chloride.

$$NaCl + H_2O + CO_2 + NH_3 \rightarrow NH_4Cl + NaHCO_3$$

During cooking

$$2NaHCO_3 \rightarrow Na_2CO_3 + H_2O + CO_2$$

Sodium hydrogen carbonate has got various uses in the household.

Uses of sodium hydrogen carbonate (NaHCO3)

(i) For making baking powder, this is a mixture of baking soda (sodium hydrogen carbonate) and a mild edible acid such as tartaric acid. When baking powder is heated or mixed in water, the following reaction takes place –

$$NaHCO_3 + H^+ \rightarrow CO_2 + H_2O + \text{Sodium salt of acid}$$

Carbon dioxide produced during the reaction causes bread or cake to rise making them soft and spongy.

(ii) Sodium hydrogen carbonate is also an ingredient in antacids. Being alkaline, it neutralizes excess acid in the stomach and provides relief.

(iii) It is also used in soda-acid fire extinguishers.

Washing soda Formation

When sodium hydrogen carbonate is heated it gives sodium carbonate. On recrystallisation i.e. adding 10 molecules of water to sodium carbonate gives washing soda.

$Na_2CO_3 + 10.H_2O \rightarrow Na_2CO_3.10H_2O$

Uses

(i) Sodium carbonate (washing soda) is used in glass, soap and paper industries.
(ii) It is used in the manufacture of sodium compounds such as borax.
(iii) Sodium carbonate can be used as a cleaning agent for domestic purposes.
(iv) It is used for removing permanent hardness of water.

Plaster of Paris

Plaster of Paris (PoP) is chemically known as Calcium Sulfate Hemihydrate. It is prepared by heating gypsum ($CaSO_4 \cdot 2H_2O$) at a very specific temperature of 373 K (100°C). At this temperature, gypsum loses three-quarters of its water of crystallization to become PoP. When you add water back to Plaster of Paris, it undergoes a hydration reaction to re-form gypsum, which sets into a very hard, rigid mass. The Chemical Reaction The equation for this transformation is:

$CaSO_4 \cdot \frac{1}{2}H_2O + 1\frac{1}{2}H_2O \rightarrow CaSO_4 \cdot 2H_2O$

Plaster of Paris + Water $\rightarrow$ Gypsum (Hard Mass)

Key Points about the Reaction: Water of Crystallization: In Plaster of Paris, two formula units of $CaSO_4$ share one molecule of water. This is why it is written as $CaSO_4 \cdot \frac{1}{2}H_2O$.

Setting Time: Once mixed with water, the paste remains workable for a short period before it begins to "set" or harden into the solid gypsum structure.

Storage: Because it reacts so readily with moisture in the air to become hard, Plaster of Paris must be stored in moisture-proof containers.

Common Uses

Because of its ability to set into a hard, molded shape, it is used extensively in various fields:

Medical: Doctors use it as a bandage for supporting fractured bones in the right position.

Construction & Decor: It is used for making smooth surfaces on walls and creating decorative ceiling designs (false ceilings).

Art: It is a popular material for making toys, statues, and casts for dental molds.

Laboratory: It can be used to seal air gaps in laboratory apparatus to make them airtight.

It is used for making toys, materials for decoration and for making surfaces smooth.

FORMATION AND USES OF COMMON SALTS

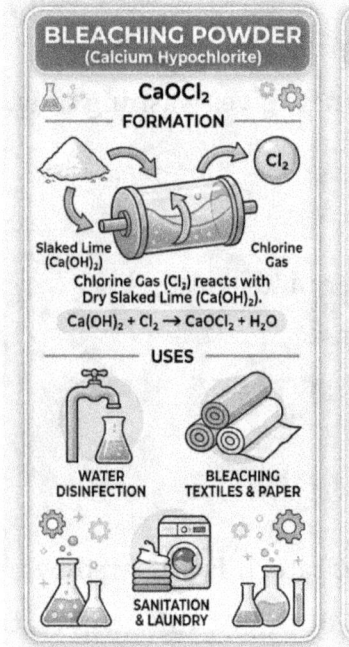
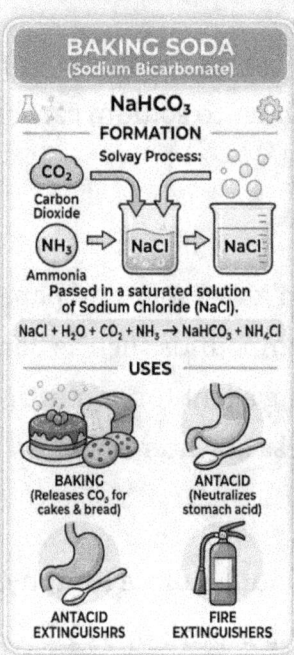
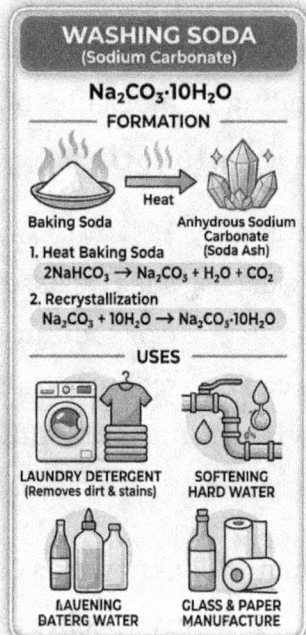
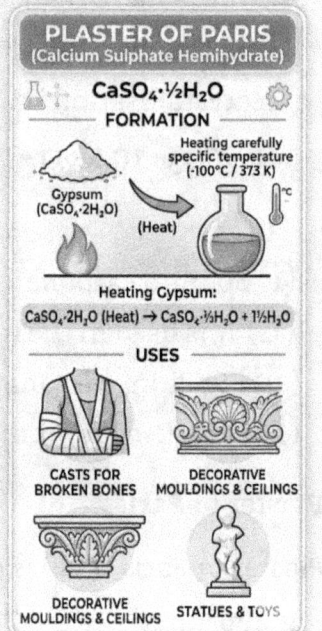

CHAPTER 6: DEEP DIVE CHALLENGES

Q1. Acid changes blue litmus to?

(a) red
(b) pink
(c) yellow
(d) no change

Q2. Phenolphthalein and methyl oranges are

(a) synthetic indicator
(b) natural indictor
(c) both
(d) none

Q3. In olfactory indicators?

(a) Colour changes
(b) Taste changes
(c) Odour changes
(d) none

Q4. Acid on reaction with metal produces?

(a) H_2
(b) O_2
(c) N_2
(d) None

Q5. Acid on reaction with metal hydrogen carbonate produces?

(a) H_2
(b) CO_2
(c) N_2
(d) None

Q6. In aqueous solution acid produces which type of ions?

(a) H^+
(b) OH^-
(c) N_2
(d) None

Q7. The process of dissolving an acid or a base in water is highly?

(a) endothermic
(b) exothermic
(c) cold
(d) None

Q8. pH scale measure which ion in a solution?

(a) H^+
(b) OH^-
(c) Cl^-
(d) None

Q9. A solution has pH value 7. It is likely to be

(a) acid
(b) base
(c) salt
(d) none of these

Q10. Our body works within pH range?

(a) 2-4
(b) 12-14
(c) 6-10
(d) 7-7.8

Q11. Milk of magnesia is?

(a) acid
(b) weak acid
(c) antacid
(d) None

Q12. Acetic acid is present in?

(a) Apple

(b) guava

(c) vinegar

(d) None

Q13. Oxalic acid is present in?

(a) potato

(b) tomato

(c) onion

(d) none

Q14. pH value of a salt is?

(a) 6

(b) 7

(c) 8

(d) 9

Q15. Formula for bleaching powder is?

(a) $CaOCl_2$

(b) $CaOCl$

(c) $CaCl_3$

(d) None

Q16. Na_2CO_3 is?

(a) Washing soda

(b) Sodium carbonate

(c) Sodium hydrogen carbonate

(d) None

Q17. Which is used for removing permanent hardness of water?

(a) $CaSO_4.1/2H_2O$

(b) $CaSO_4.2H_2O$

(c) $CaSO_4$

(d) Washing soda

Q18. A solution turns blue litmus red its pH is likely to be?

(a) 1

(b) 7

(c) 8

(d) 9

Q19. A solution reacts with crushed egg-shells to give a gas that turns lime-water milky. The solution contains

(a) NaCl

(b) HCl

(c) LiCl

(d) KCl

Q20. Which one of the following types of medicines is used for treating indigestion?

(a) Antibiotic
(b) Analgesic
(c) Antacid
(d) Antiseptic

Match the following:

Q21.

A	B
Acetic acid	Nettle sting
Citric acid	Curd
Lactic acid	Vinegar
Methanoic acid	Orange

Q22.

A	B
Bleaching powder	$CaSO_4.2H_2O$
Baking soda	$CaSO_4.1/2H_2O$
Washing soda	$CaOCl_2$
Plaster of paris	$Na_2CO_3.10H_2O$
Gypsum	$NaHCO_3$

Q23.

A	B
Acid	Red cabbage
Base	Salt
pH	H^+
Natural indicator	OH^-

Short Answer Questions

Q24. Write down physical differences between acid and bases.

Q25. Write down chemical differences between acid and bases.

Q26. What are common properties between acid and base?

Q27. What is salt and how it is formed?

Q28. What is anta-acids? How they help us in digestion?

Q29. Write down the uses of baking soda?

Q30. How does plaster of Paris is formed?

Q31. Why should curd and sour substances not be kept in brass and copper vessels?

Q32. Why does dry HCl gas not change the color of the dry litmus paper?

Long answer type questions:

Q33. Suppose you are sent to a planet where total soil is acidic in nature. It is also not suitable for agriculture purpose. You have to grow plant there. What procedure you can use to solve this problem?

Q34. Write down the formation and uses of following compounds:

(a) Washing soda

(b) Baking soda

(c) Bleaching powder

(d) Plaster of paris

Q35. Write down five chemical properties of acids. Explai with chemical equation in each case.

Q36. Write down five chemical properties of bases. Explai with chemical equation in each case.

Solution

1.a 2.a 3.c 4.a 5.b 6.a 7.b 8.a 9.c 10.d 11.c 12.c 13.b 14.b 15.a 16.b 17.d 18.a 19.b 20.c

Short Answer Hints

- Q24 (Physical): Think about taste (sour vs. bitter) and touch (one can be corrosive, the other feels soapy).
- Q25 (Chemical): Focus on the specific ions they release in water (H^+ vs. OH^-) and how they change the color of Litmus paper.
- Q26 (Commonalities): Think about their ability to conduct electricity in solution and their shared goal in a neutralization reaction.
- Q27 (Salt): It's the "offspring" of a chemical marriage. What happens when an acid and a base cancel each other out?
- Q28 (Antacids): These are mild bases. If your stomach has too much "fire" (acid), how does a base help cool it down?

BASIC CHEMISTRY

- Q29 (Baking Soda): Think of fluffy cakes (releasing CO2), fire extinguishers, and its use as a mild antiseptic.
- Q30 (P.O.P.): It comes from a mineral called Gypsum. What happens when you carefully heat Gypsum at exactly 373 K?
- Q31 (Vessels): Sour substances contain acids. What happens when an acid reacts with a metal like Copper? Is the resulting product safe to eat?
- Q32 (Dry HCl): Acids only show their "true colors" (ionize) in the presence of water. If everything is dry, can the H^+ ions move?

Long Answer Hints

Q33: The Acidic Planet
- The Problem: The soil pH is too low for plants.
- The Solution: Think about Neutralization. What "household" or industrial basic substances (like Lime or Chalk) could you add to the soil to raise the pH?

Q34: Formation and Uses
- (a) Washing Soda: Look up the Solvay Process and the recrystallization of Sodium Carbonate. (Use: Laundry/Glass making).
- (b) Baking Soda: Reaction of Brine with CO2 and NH3. (Use: Antacid/Baking).
- (c) Bleaching Powder: Action of Chlorine gas on Dry Slaked Lime. (Use: Disinfecting water).
- (d) Plaster of Paris: Half-hydration of Calcium Sulphate. (Use: Fixing fractured bones).

Q35: Chemical Properties of Acids
Think about what happens when an acid meets:
1. Metal: (Releases Hydrogen gas).
2. Metal Carbonate: (Releases Carbon Dioxide).
3. Base: (Forms Salt and Water).
4. Metal Oxide: (Basic in nature, so think neutralization).
5. Water: (Heat is released; always add acid to water, not water to acid!).

Q36: Chemical Properties of Bases
Think about what happens when a base meets:
1. Certain Metals: (Like Zinc or Aluminum—releases Hydrogen).
2. Acid: (Neutralization).
3. Non-Metal Oxide: (Non-metal oxides are acidic, so what is the result?).
4. Ammonium Salts: (Often releases Ammonia gas).
5. Water: (Dissolving to form Alkalis).

CHAPTER 7

Metals and Non-Metals

7.1 Introduction

Physical properties of metals and non-metals

7.1 HARDNESS

Hardness refers to strength of a material. Generally, metals are hard and non-metals are soft and brittle. There are some exceptions, lead is a brittle metal. Some metals are so soft that they can be cut by knife such as sodium, potassium, magnesium.

7.2 MALLEABILITY

Malleability is the property of being beaten in the form of a thin sheet. Metals can be beaten in the thin sheets. This property helps us to use metals sheets, containers etc.

7.3 DUCTILITY

Metals can be drawn into the wires this property is known as ductility. Gold is most ductile element that's why ornaments are not form from the pure gold.

7.4 CONDUCTION OF HEAT

The materials which allow heat to pass through them are known as the conductor of heat. Metals are good conductor of heat and nonmetals are bad.

7.5 CONDUCTION OF ELECTRICITY

The materials which allow electricity to pass through them are known as the conductor of electricity. Metals are good conductors and nonmetals are insulator. Graphite is a nonmetal but a good conductor of electricity.

7.6 SONORITY

Sonority refers to the sounds produced when a metal is beaten. Due to this property copper or irons are used as a bell.

7.7 METALLIC LUSTER

Metals have a shining surface while nonmetals have not. This property is referred as metallic luster. Iodine is exception; it is lustrous but nonmetal.

COMPARATIVE PHYSICAL PROPERTIES: METALS vs. NON-METALS

METALS		NON-METALS
GENERALLY HARD, DENSE (Except Na, K)	HARDNESS	GENERALLY SOFT, LOW DENSITY (Diamond is hard)
CAN BE HAMMERED INTO SHEETS	MALLEABILITY	BRITTLE, SHATTER WHEN STRUCK
CAN BE DRAWN INTO WIRES	DUCTILITY	NOT DUCTILE
GOOD CONDUCTORS OF HEAT	HEAT CONDUCTION	POOR CONDUCTORS OF HEAT
GOOD CONDUCTORS OF ELECTRICITY (Free electrons)	ELECTRICITY CONDUCTION	POOR CONDUCTORS (Graphite is an exception)
SONOROUS (Ring when struck)	SONORITY	NOT SONOROUS
SHINY AND LUSTROUS	METALLIC LUSTRE	GENERALLY DULL

*Exceptions apply in some cases

7.8 CHEMICAL PROPERTIES OF METALS

In the previous lesson we have seen the reaction of metals and non-metals with acids, bases, carbonates, hydrogen carbonates, and oxides. In this section we will see what happens when metals and a non-metal react.

7.9 REACTION WITH SOLUTION OF OTHER SALTS

When metals react with solution of the other salts displacement reaction occurs. About we have already discussed in the chemical reaction part. The following reactivity series will help in finding the more reactive and less reactive elements.

7.10 THE REACTIVITY SERIES

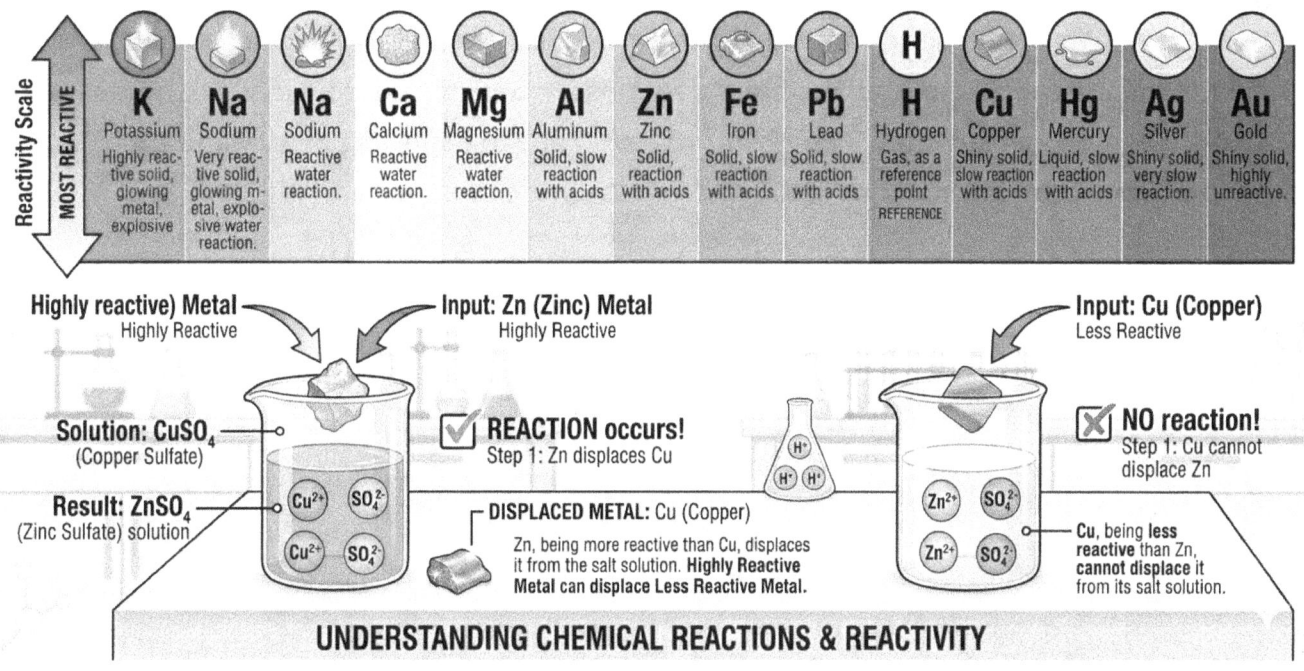

7.11 REACTION WITH NON-METALS

In the first part of this series, in chapter 0 (some basic concepts) we have studied that when metals and nonmetals come into contact electrons are transferred from metal to nonmetal and anions and cations are formed. These are attracted due to opposite polarity and a strong ionic bond is formed. These compounds are referred as ionic compounds.

7.12 IONIC COMPOUNDS AND ITS PROPERTIES

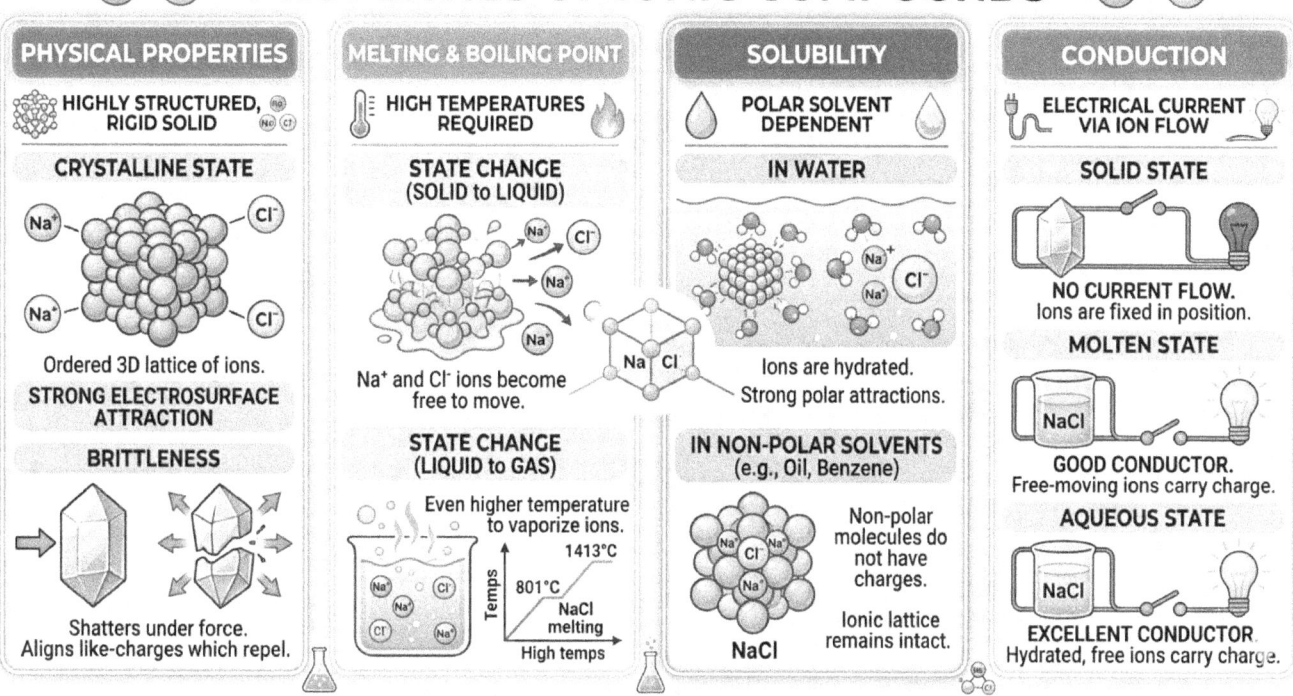

(i) *Physical nature*: Ionic compounds are solids and are somewhat hard because of the strong force of attraction between the positive and negative ions. These compounds are generally brittle and break into pieces when pressure is applied.

(ii) *Melting and Boiling points*: Ionic compounds have high melting and boiling points. This is because a considerable amount of energy is required to break the strong inter-ionic attraction.

(iii) *Solubility*: Electrovalent compounds are generally soluble in water and insoluble in solvents such as kerosene, petrol, etc.

(iv) *Conduction of Electricity*: The conduction of electricity through a solution involves the movement of charged particles. A solution of an ionic compound in water contains ions, which move to the opposite electrodes when electricity is passed through the solution. Ionic compounds in the solid state do not conduct

electricity because movement of ions in the solid is not possible due to their rigid structure.

But ionic compounds conduct electricity in the molten state. This is possible in the molten state since the electrostatic forces of attraction between the oppositely charged ions are overcome due to the heat. Thus, the ions move freely and conduct electricity.

7.13 ORES AND MINERALS

Compounds or elements which naturally occur in the earth's crust are known as ***minerals***. The best example is water. Profitable minerals are referred as ***ores***. If in getting a mineral we are in loss economically it's not an ore. The unwanted sand soil and mud contaminated with ore are referred as gangue.

7.14 EXTRACTION OF METALS

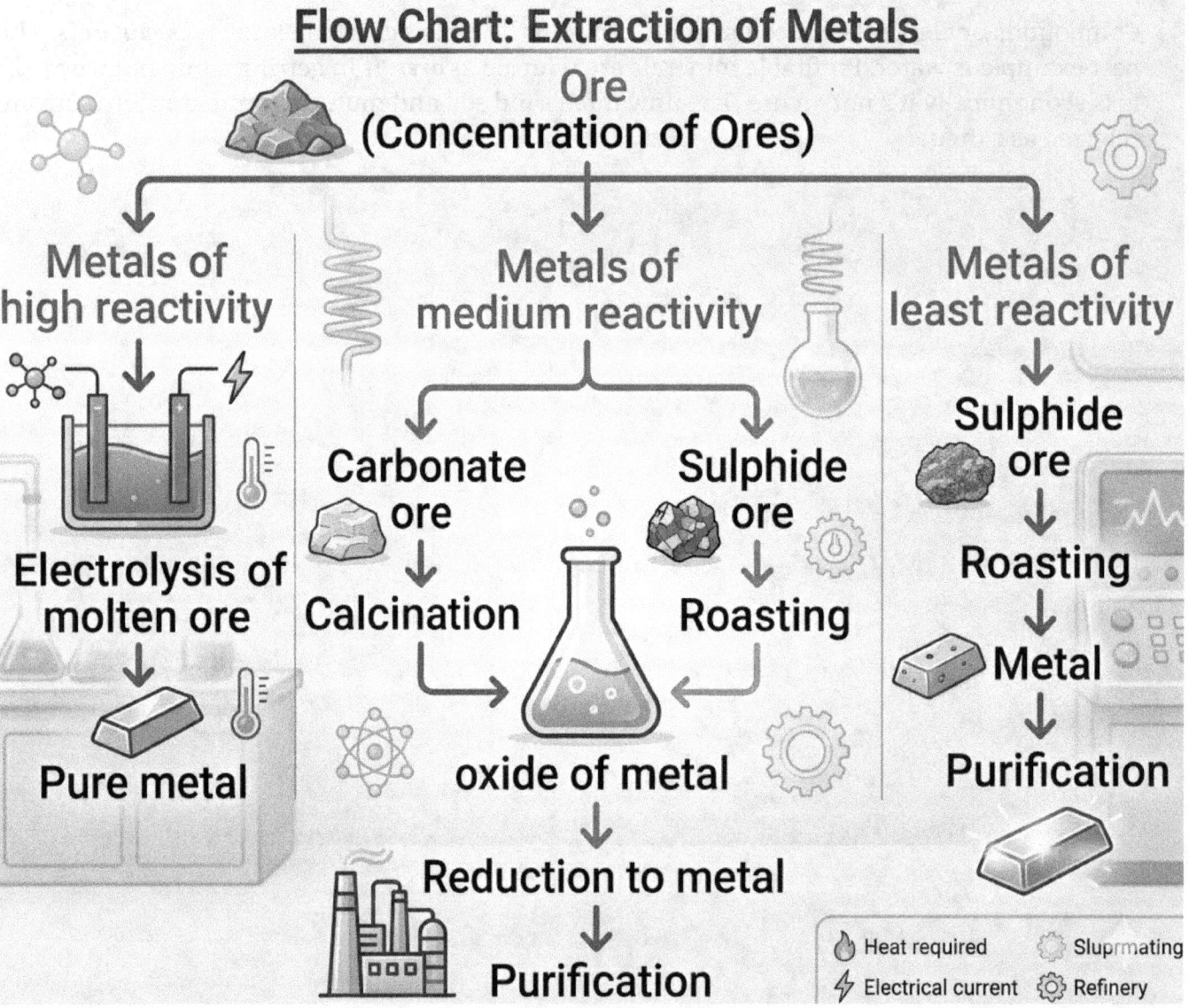

7.15 ENRICHMENT OF ORES

Ores mined from the earth are usually contaminated with large amounts of impurities such as soil, sand, etc., called *gangue*. The impurities must be removed from the ore prior to the extraction of the metal. The processes several steps are involved in the extraction of pure metal from ores. A summary of these steps is given in flow chart. Each step is explained in detail in the following sections.

The processes used for removing the gangue from the ore are based on the differences between the physical or chemical properties of the gangue and the ore. Different separation techniques are accordingly employed.

7.16 EXTRACTING METALS LOW IN THE REACTIVITY SERIES

Metals lows in the activity series are very unreactive. The oxides of these metals can be reduced to metals by heating alone. For example, cinnabar (HgS) is an ore of mercury. When it is heated in air, it is first converted into mercuric oxide (HgO). Mercuric oxide is then reduced to mercury on further heating.

$2HgS(s) + 3O(g) \rightarrow 2HgO(s) + 2SO_2(g)$

$2HgO(s) \rightarrow 2Hg(l) + O_2(g)$

Similarly, copper which is found as Cu_2S in nature can be obtained from its ore by just heating in air.

$2CuS + 3O(g) \rightarrow 2CuO(s) + 2SO(g)$

$2CuO + 2CuS \rightarrow 4Cu + 2SO_2$

7.17 EXTRACTING METALS IN THE MIDDLE OF THE REACTIVITY SERIES

The metals in the middle of the activity series such as iron, zinc, lead, copper, etc., are moderately reactive. These are usually present as sulphides or carbonates in nature. It is easier to obtain a metal from its oxide, as compared to its sulphides and carbonates. Therefore, prior to reduction, the metal sulphides and carbonates must be converted into metal oxides. The sulphide ores are converted into oxides by heating strongly in the presence of excess air. This process is known as *roasting*.

The carbonate ores are changed into oxides by heating strongly in limited air. This process is known as *calcination*. The chemical reaction that takes place during roasting and calcination of zinc ores can be shown as follows –

Roasting
$$2ZnS(s) + 3O(g) \rightarrow 2ZnO(s) + 2SO_2(g)$$

Calcination
$$ZnCO_3 \rightarrow (s) ZnO(s) + CO_2(g)$$

The metal oxides are then reduced to the corresponding metals by using suitable reducing agents such as carbon.
For example,
when zinc oxide is heated with carbon, it is reduced to metallic zinc.

$$ZnO(s) + C(s) \rightarrow Zn(s) + CO(g)$$

Besides using carbon (coke) to reduce metal oxides to metals, sometimes displacement reactions can also be used. The highly reactive metals such as sodium, calcium, aluminium, etc., are used as reducing agents because they can displace metals of lower reactivity from their compounds. For example, when manganese dioxide is heated with aluminium powder, the following reaction takes place –

$$3MnO_2(s) + 4Al(s) \rightarrow 3Mn(l) + 2Al_2O_3(s) + Heat$$

These displacement reactions are highly exothermic. The amount of heat evolved is so large that the metals are produced in the molten state. In fact, the reaction of iron (III) oxide (Fe_2O_3) with aluminum is used to join railway tracks or cracked machine parts. This reaction is known as the *thermit reaction*.

$Fe_2O_3(s) + 2Al(s) \rightarrow 2Fe(l) + Al_2O_3(s) + Heat$

7.18 EXTRACTING METALS TOWARDS THE TOP OF THE REACTIVITY SERIES

The metals high up in the reactivity series are very reactive. They cannot be obtained from their compounds by heating with carbon. For example, carbon cannot reduce the oxides of sodium, magnesium, calcium, aluminium, etc., to the respective metals. This is because these metals have more affinity for oxygen than carbon. These metals are obtained by electrolytic reduction. For example,

sodium, magnesium and calcium are obtained by the electrolysis of their molten chlorides. The metals are deposited at the cathode (the negatively charged electrode), whereas, chlorine is liberated at the anode (the positively charged electrode). The reactions are –

At cathode $Na^+ + e^- \rightarrow Na$ At
anode $2Cl^- \rightarrow Cl_2 + 2e^-$

Similarly, aluminium is obtained by the electrolytic reduction of aluminium oxide.

7.19 REFINING OF METALS

The metals produced by various reduction processes described above are not very pure. They contain impurities, which must be removed to obtain pure metals. The most widely used method for refining impure metals is electrolytic refining.

Electrolytic Refining: Many metals, such as copper, zinc, tin, nickel, silver, gold, etc., are refined electrolytically. In this process, the impure metal is made the anode and a thin strip of pure metal is made the cathode. A solution of the metal salt is used as an electrolyte. On passing the current through the electrolyte, the pure metal from the anode dissolves into the electrolyte. An equivalent amount of pure metal from the electrolyte is deposited on the cathode. The soluble impurities go into the solution, whereas, the insoluble impurities settle down at the bottom of the anode and are known as *anode mud*.

CHAPTER 7: DEEP DIVE CHALLENGES

Q1. Which metal is liquid at room temperature?

(a) Hg
(b) Br
(c) Cl
(d) None

Q2. Which nonmetal is liquid at room temperature?

(a) Hg
(b) Br
(c) Cl
(d) None

Q3. Which non metal is good conductor of electricity?

(a) Cu
(b) Fe
(c) Graphite
(d) Diamond

Q4. Which of the following can be cut by knife?

(a) Na
(b) Mg
(c) K
(d) All of the above

Q5. Calcination is used for?

(a) Sulphide ore
(b) Carbonate ore
(c) Oxide ore
(d) None of these

Q6. Which one is more reactive?

(a) Cu
(b) Pb
(c) Fe
(d) Zn

Q7. In thermite reaction the product is?

(a) Fe(l)
(b) Al_2O_3
(c) heat
(d) All of the above

Q8. In galvanization the layer of which metal is coated on iron or steel articles?

(a) Zn
(b) Mg
(c) K
(d) none of these

Q9. An alloy is a homogeneous mixture of?

(a) Two or more metals
(b) Metal and non-metals
(c) Both a and b
(d) None of these

Q10. Al_2O_3 is?

(a) Metallic oxide
(b) Nonmetallic oxides
(c) Amphoteric oxide
(d) None

Q11. Which types of substances are effective in cleaning the vessels?

(a) Sour
(b) Oily
(c) Sweets
(d) None of these

Q12. In electrolytic refining the pure metal is taken as?

(a) Anode
(b) Cathode
(c) Anyone can be taken

(d) None of these

Q13. Copper is used to make hot water tank because?

(a) It is good conductor of heat
(b) It is good conductor of electricity
(c) malleability
(d) Ductility

Q14. Coins are manufactured from silver and copper because?

(a) These are good conductor of heat
(b) These are good conductor of electricity
(c) Malleability
(d) Ductility

Q15. Nonmetallic oxides are?

(a) Acidic
(b) Neutral
(c) Either acidic or neutral
(d) None of these

Q16. Alloy of Cu and Zn is?

(a) Bronze
(b) Solder
(c) Brass
(d) Amalgam

Q17. In amalgam which element must be present?

(a) Hg
(b) Pb
(c) Fe
(d) Zn

Q18. solder has?

(a) High melting point
(b) Low melting point
(c) Moderate melting point
(d) Extremely low melting point

Q19. cinebar is Ore of?
- (a) Hg
- (b) Pb
- (c) Fe
- (d) None of these

Q20. Copper pyrite is ore of?
- (a) Cu
- (b) Pb
- (c) Fe
- (d) Zn

Match the following:

Q21.

A	B
Amphoteric	CO_2
Acidic	Al_2O_3
Basic	NaOH
Neutral	H_2O

Q22.

A	B
Ionic	CH_4
Co-valent	NaCl
Roasting	$ZnCO_3$
Calcination	ZnS

Q23.

A	B
Solder	Zn & Cu
Brass	Cu & Sn
Bronze	Hg & Al
Amalgam	Pb & Sn

Short Answer Questions

Q24. What are physical differences between metal and nonmetals?
Q25. What are chemical differences between metal and nonmetals?
Q26. What are alloys? Give examples.
Q27. What is difference between ores and minerals?
Q28. What are amphoteric oxides?
Q29. What is thermite reaction?
Q30. What happens when zinc reacts with cold water and steam?
Q31. Which type of compound is formed when metal reacts with nonmetals?
Q32. What is difference between calcinations and roasting?
Q33. How does metals are extracted from its oxides?

Long answer type questions

Q34. Give reasons

(a) Platinum, gold and silver are used to make jewellery.

(b) Sodium, potassium and lithium are stored under oil.

(c) Aluminium is a highly reactive metal, yet it is used to make utensils for cooking.

(d) Carbonate and sulphide ores are usually converted into oxides during the process of extraction.

Q35. Pratyush took sulphur powder on a spatula and heated it. He collected the gas evolved by inverting a test tube over it.
What will be the action of gas on:

(a) Dry litmus paper?

(b) Moist litmus paper?

(c) Write a balanced chemical equation for the reaction taking place.

Solution

1.a 2.b 3.c 4.d 5.b 6.d 7.d 8.a 9.c 10.c 11.a 12.b 13.a 14.c 15.c 16.c 17.a 18.b 19.a 20.a

Short Answer Hints
- Q24 (Physical): Think about "The 3 S's": Shininess (lustre), Sonority (ringing sound), and Strength (malleability/ductility). Also, consider which one is a better "highway" for heat and electricity.
- Q25 (Chemical): Focus on the oxides. When they react with oxygen, which one forms a basic oxide and which one forms an acidic one? Also, think about which loses electrons (+) and which gains them (-).
- Q26 (Alloys): It's a "mixture" of a metal with something else to make it stronger or rust-proof. Think of Steel, Brass, or the "Gold" used in jewelry.
- Q27 (Ores vs. Minerals): All ores are minerals, but not all minerals are ores. The key is profit—which one can you get the metal out of easily and cheaply?
- Q28 (Amphoteric): These are the "dual-natured" oxides. They can act like an acid when

- meeting a base, and like a base when meeting an acid. (Example: Aluminum or Zinc oxides).
- Q29 (Thermite): A very violent reaction used to join railway tracks. It involves Aluminum stealing oxygen from Iron oxide, releasing massive heat.
- Q30 (Zinc & Water): Zinc is "picky." It is too lazy to react with cold water. It needs more energy—does it react with hot water or only with the gas form (steam)?
- Q31 (Compounds): When a metal (giver) meets a non-metal (taker), they form a bond based on attraction. Think of table salt (NaCl).
- Q32 (Calcination vs. Roasting): One is for Carbonate ores (think "C" for Calcination) and happens in limited air. The other is for Sulphide ores and needs "Roasting" in plenty of air.
- Q33 (Extraction): To get the metal, you must remove the Oxygen. This process is called Reduction. Think about using Carbon (Coke) or Electricity.

Long Answer Hints

Q34: Give Reasons
- (a) Jewelry: Why would you want a ring that doesn't tarnish or react with air and water even after many years?
- (b) Storage in Oil: These metals are "hyperactive." If they touch air or even a tiny drop of moisture, they might catch fire. Oil keeps them "sleeping."
- (c) Aluminum Utensils: Even though it's reactive, it quickly forms a protective invisible layer of oxide on its skin that stops further damage.
- (d) Conversion to Oxides: It is much easier to "pull" a metal out of an Oxide than it is to pull it out of a Carbonate or Sulphide.

Q35: The Sulphur Experiment
- The Gas: Heating Sulphur creates Sulphur Dioxide SO2).
- (a) Dry Litmus: Does a gas show its acidic nature without water to create ions?
- (b) Moist Litmus: When SO_2 meets the water on the paper, it forms Sulphurous Acid. What color do acids turn litmus?
- (c) Equation: $S + O_2 \rightarrow ?$ and then $SO_2 + H_2O \rightarrow ?$

CHAPTER 8

Carbon and its compounds

8.1 INTRODUCTION

Carbon is present in almost every matter around us. Maximum things around us on burning changed to the ashes (oxides of carbon). For example: the sugar we eat, cloths we wear and food items we eat all produces carbon on burning. It is such a versatile and Omni present element; it forms maximum no. of compounds in the world.

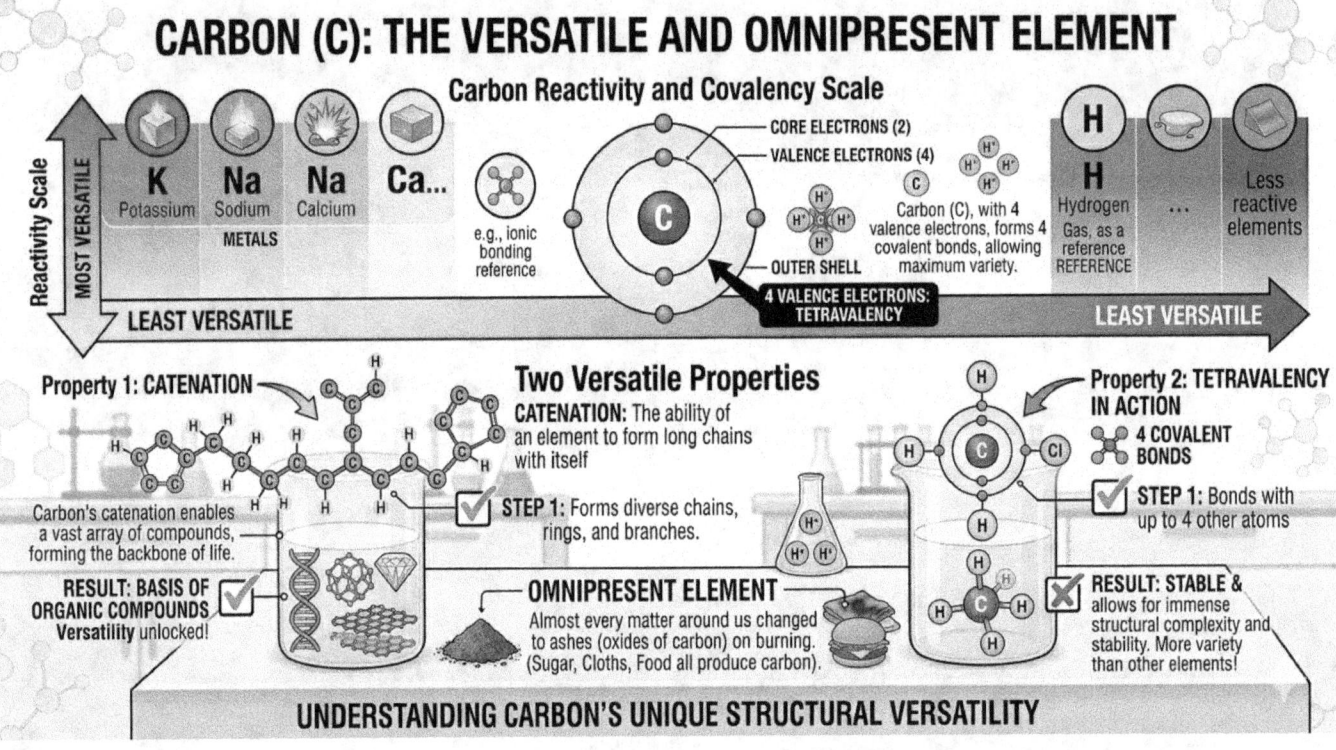

8.2 ABOUT CARBON

As we have already studied its atomic no. is 6 and its atomic mass is 12 amu. It has 6 protons and 6 electrons in a C-12 carbon atom. During reaction whether it can take 4 electrons or lose 4 electrons in order to get a stable configuration. In first case 6 protons have to balance 10 electrons which seems unstable, in second case it requires much more energy to release 4 electrons. To overcome this problem carbon shares, its electrons with other elements or same type carbon atoms.

The compounds which are formed from the combination of hydrogen and carbon are generally referred as hydrocarbons. However, these compounds can have some other elements like sulphur, oxygen, nitrogen etc as a functional group. About functional group we will study in sub sequent sections.

8.3 COVALENT BOND

We have already studied in part 1, about covalent bonding. Here we will see that in covalent bonding the shared electrons are counted in both atoms to full fill the stable configuration. These bonding also can be expressed in the form of electron dot structure. Each electron is represented as a dot. The combination of two dots forms a single bond.

8.4 VERSATILE NATURE OF CARBON

There are two versatile properties of carbon through which it forms infinite no. of carbon compounds. These are as follows:

1. Its four valency which provides a lot of branching.

2. Its property to link with same type of atoms generally known as *catenation*.

8.5 SATURATED AND UNSATURATED CARBON COMPOUNDS

Further hydrocarbons can be divided in two categories i.e. saturated and unsaturated. Saturated hydrocarbons contain single bonds while unsaturated contain double or triple bonds. There other differences also between them, for example: saturated hydrocarbons give blue flame on complete combustion while unsaturated give a yellow sooty flame. Saturated hydrocarbons undergo substitution reaction while unsaturated can't. Generally saturated are fairly inert in the presence of most of the reagents while unsaturated are not. Alkanes are saturated while alkenes and alkynes are unsaturated in nature.

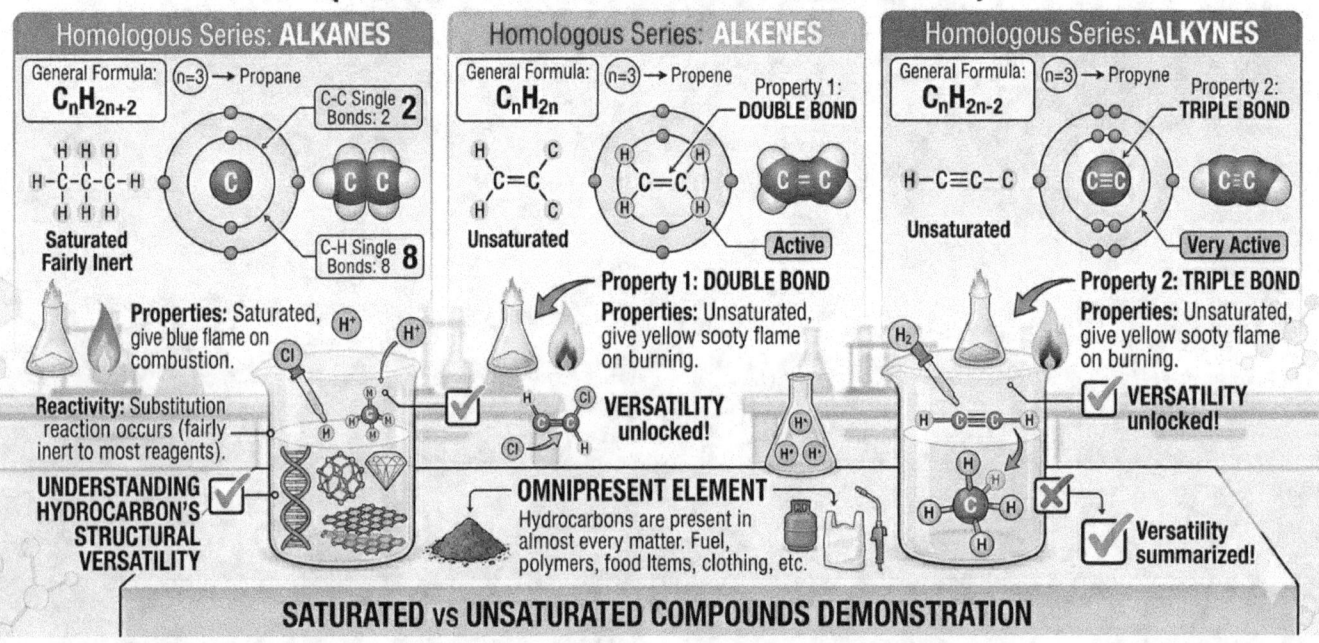

8.6 ALKANE

Alkanes have general formula C_nH_{2n+2}, where n refers to the no. of atoms. This formula shows that the no. hydrogen atom is two more than the double of the carbon atom. Meth, eth, prop but, pent, hex, hept, oct, non and dec signifies the no. of carbon atom 1, 2, 3, 4 5, 6, 7, 8, 9 and 10 respectively while –ane is joined to these root words to form the name of the compounds. The following table shows the names of first ten alkanes.

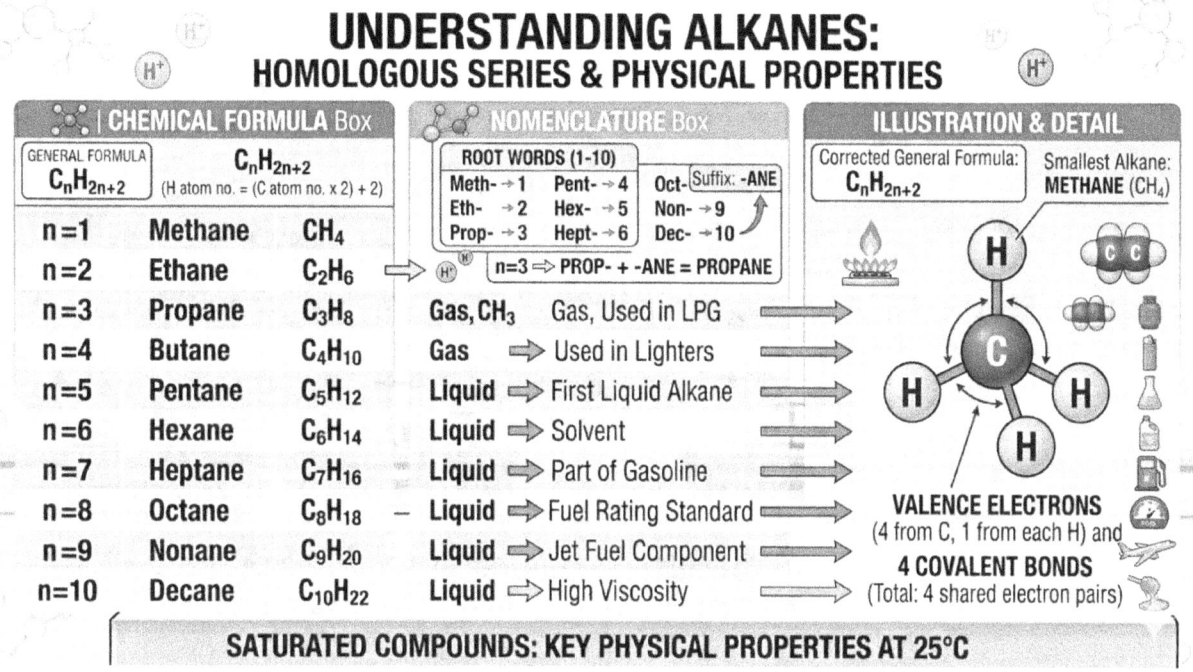

8.7 ALKENE

Alkenes have general formula C_nH_{2n}, where n refers to the no. of atoms. This formula shows that the no. hydrogen atom is double of the carbon atom. In this compound there must be one double bond.

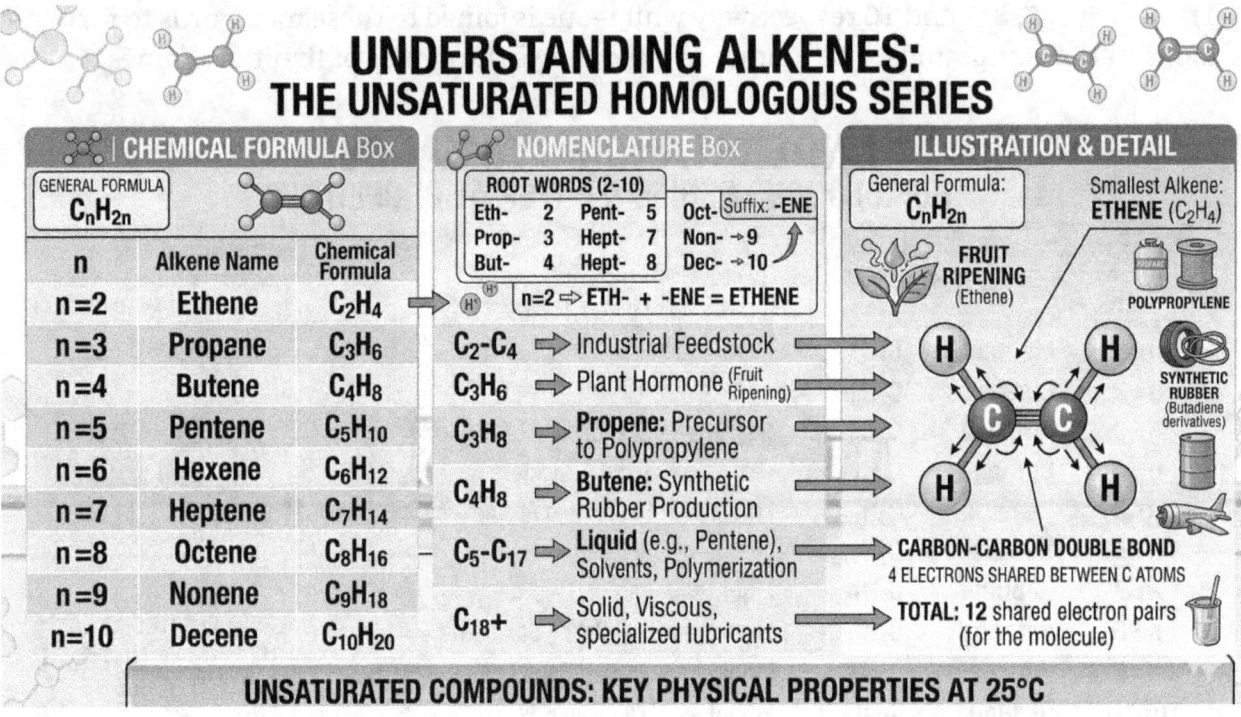

8.8 ALKYNE

Alkynes have general formula C_nH_{2n-2}, where n refers to the no. of atoms. This formula shows that the no. hydrogen atom is two less than the double of the carbon atom. In this compound there must be one triple bond.

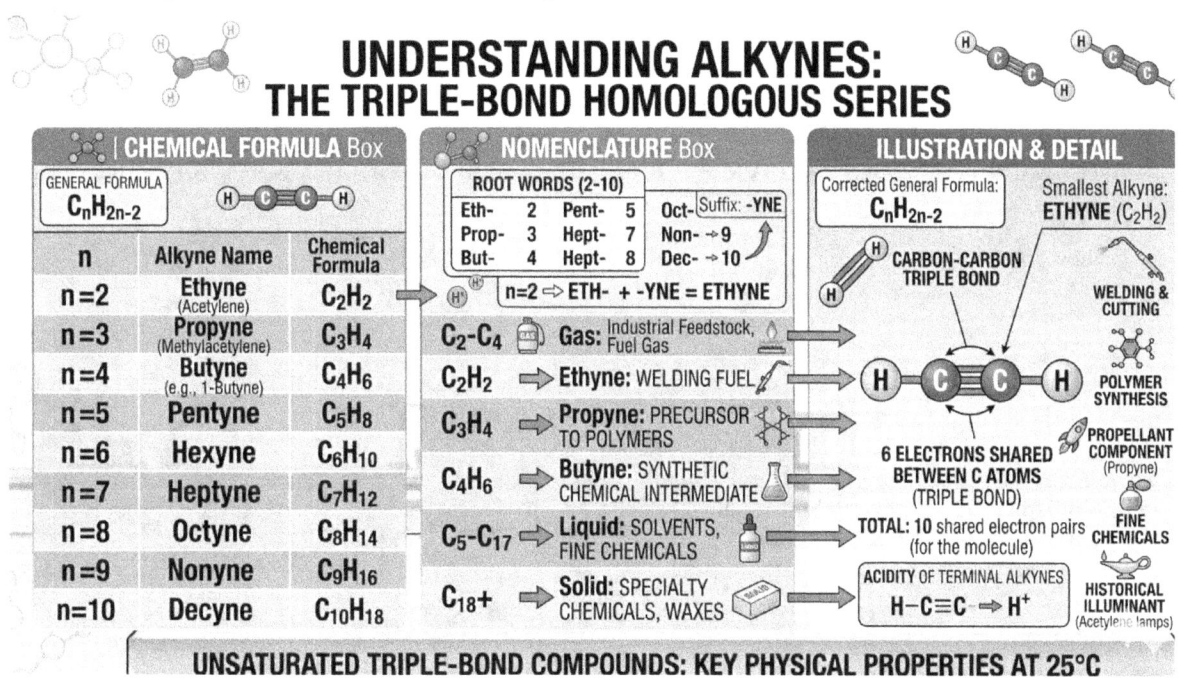

8.9 FUNCTIONAL GROUPS

When one hydrogen atom is removed from the alkane, a new group is formed this is known as **alkyl group**. It is generally denoted by –R. Functional group is the atom or group of atoms which when attached with alkyl group can change its physical and chemical properties.

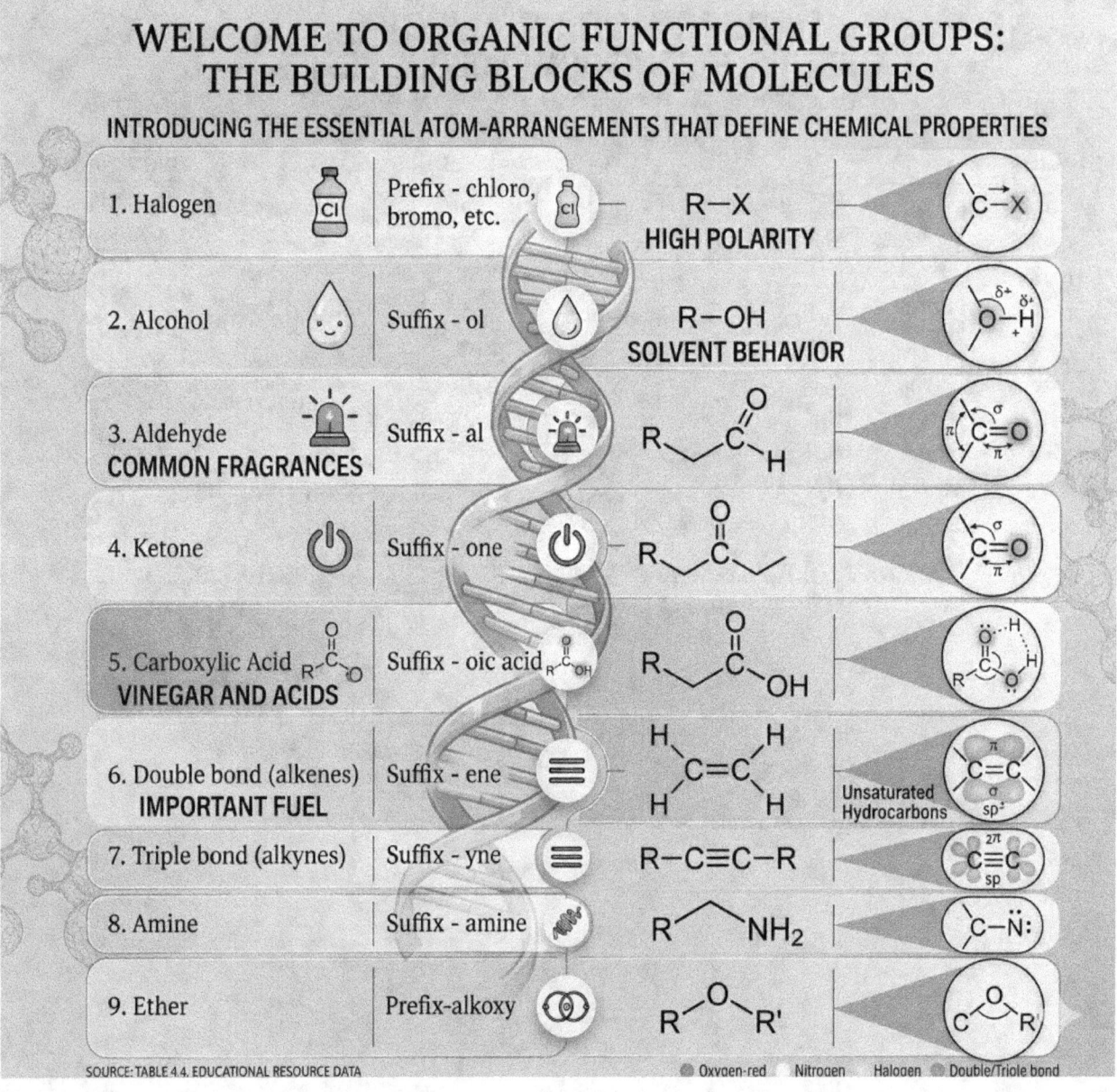

8.10 HALO- GROUP

These are the halogen atoms, such as fluorine, chlorine, bromine, iodine etc.

8.11 ALCOHOL GROUP

This group consists of hydrogen and oxygen i.e. –OH.

8.12 ALDEHYDE GROUP

Aldehyde group consists of C, H and O in the following fashion:

8.13 KETONE GROUP

This group is –CO. in this group there is double bond between carbon and oxygen. And carbon is attached with at least two alkyl group.

8.14 ALDEHYDE AND KETONE

However, this double bond between carbon and oxygen also present in the aldehyde group but it is ketone only in which carbon atom of functional group is attached to at least two alkyl groups.

8.15 CARBOXYLIC ACID

Carboxylic acid has the functional group –COOH. There is one double bond present between the carbon and oxygen, while there is single bond between hydrogen and oxygen.

8.16 NOMENCLATURE OF CARBON COMPOUNDS

Naming a carbon compound can be done by the following method –

(i) Identify the number of carbon atoms in the compound. A compound having three carbon atoms would have the name propane.

(ii) In case a functional group is present, it is indicated in the name of the compound with either a prefix or a suffix (as given in Table 4.4).

(iii) If the name of the functional group is to be given as a suffix, the name of the carbon chain is modified by deleting the final 'e' and adding the appropriate suffix. For example, a three-carbon chain with a ketone group would be named in the following manner – Propane – 'e' = propan + 'one' = propanone.

(iv) If the carbon chain is unsaturated, then the final 'ane' in the name of the carbon chain is substituted by 'ene' or 'yne' as given in Table 4.4. For example, a three-carbon chain with a double bond would be called propene and if it has a triple bond, it would be called propyne.

THE COMPREHENSIVE GUIDE TO ORGANIC FUNCTIONAL GROUP NOMENCLATURE: 9 ESSENTIAL CLASSES

FUNCTIONAL GROUP	PREFIX/SUFFIX	CHEMICAL STRUCTURE	MOLECULE NAME
1. Halogen	Prefix - chloro, bromo, etc.	H-C-C-C-Cl (with H's)	Chloropropane / Bromopropane
2. Alcohol	Suffix - ol	H-C-C-C-OH (with H's)	Propanol
3. Aldehyde	Suffix - al	H-C-C-C=O (with H's)	Propanal
4. Ketone	Suffix - one	H-C-C(=O)-C-H (with H's)	Propanone
5. Carboxylic acid	Suffix - oic acid	H-C-C-C(=O)-OH (with H's)	Propanoic acid
6. Double bond (alkenes)	Suffix - ene	H-C-C=C-H (with H's)	Propene
7. Triple bond (alkynes)	Suffix - yne	H-C-C≡C-H	Propyne
8. Amine	Suffix - amine	$H_3C-CH_2-CH_2-NH_2$	Propan-1-amine
9. Ether	Prefix - alkoxy	$CH_3-CH_2-O-CH_2-CH_3$	Ethoxyethane

SOURCE: TABLE 4.4, EDUCATIONAL RESOURCE DATA

O Cl N N Triple bond etc

8.17 CHEMICAL PROPERTIES OF CARBON COMPOUNDS

Combustion: In combustion reaction the hydrocarbons on combustion produces heat and light.

(i) $C + O_2 \rightarrow CO_2$ + heat and light

(ii) $CH_4 + O_2 \rightarrow CO_2 + H_2O$ + heat and light

(iii) $CH_3CH_2OH + O_2 \rightarrow CO_2 + H_2O$ + heat and light

Oxidation: oxidation refers to the addition of oxygen. This reaction is generally used to convert alcohols to carboxylic groups. When we add oxygen to alcohol using an oxidizing agent like alkaline $KMnO_4$ or acidified $K_2Cr_2O_7$, alcohol changes to carboxylic acid.

Addition reaction: addition reaction is used to convert unsaturated hydrocarbon into saturated hydrocarbon. In this reaction a single bond is broken to associate with halogens. In many cases in place of halogen H_2 is used with nickel as a

catalyst. it is commercially used to convert vegetable oil into animal fats.

Substitution reaction: Saturated hydrocarbons are fairly unreactive and are inert in the presence of most reagents. However, in the presence of sunlight, chlorine is added to hydrocarbons in a very fast reaction. Chlorine can replace the hydrogen atoms one by one. It is called a substitution reaction because one type of atom or a group of atoms takes the place of another. A number of products are usually

formed with the higher homologues of alkanes.

$CH_4 + Cl_2 \rightarrow CH_3Cl + HCl$ (in the presence of sunlight)

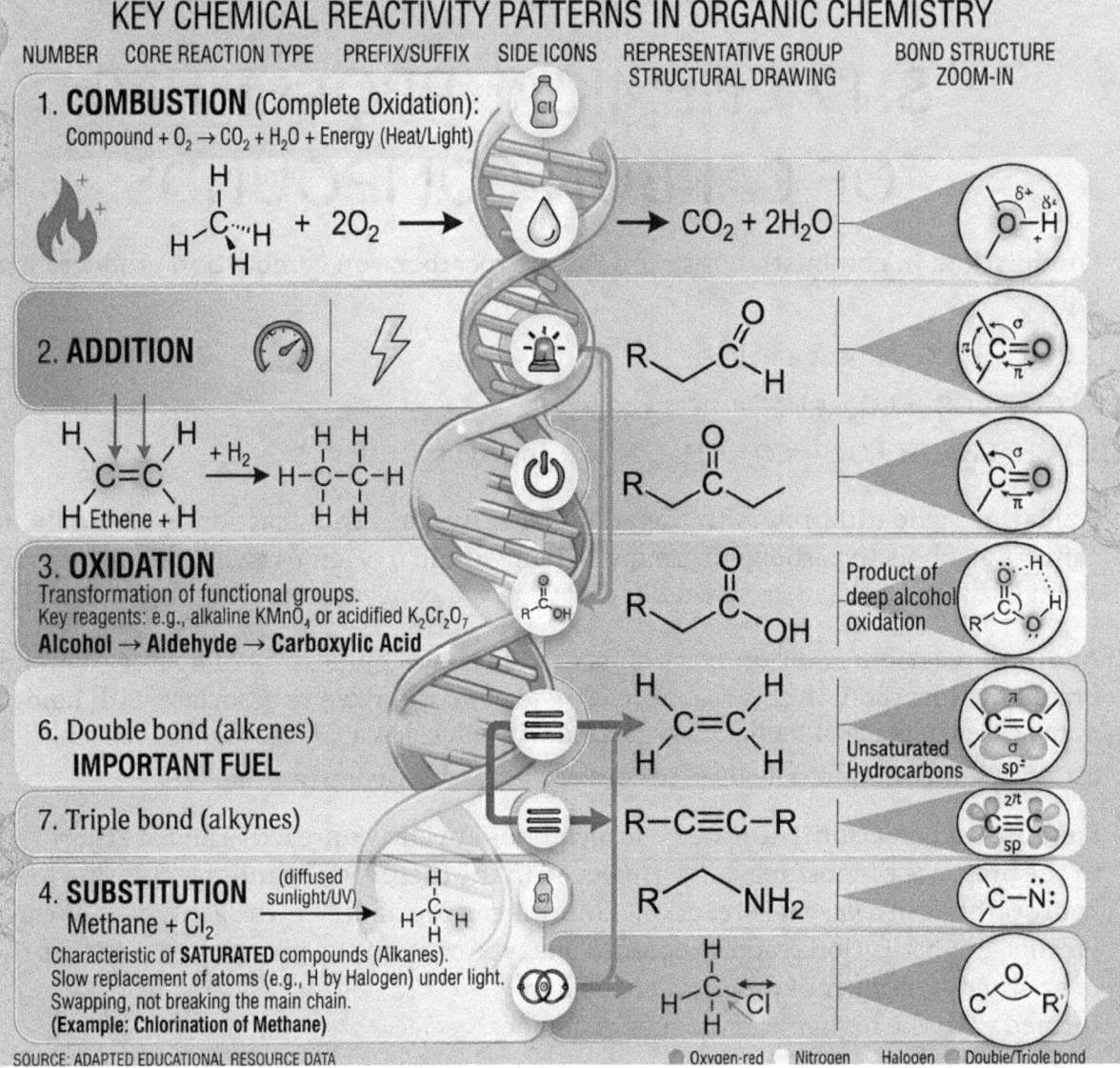

8.18 SOME IMPORTANT CARBON COMPOUNDS

8.19 ETHANOL

Physical properties

Ethanol is a liquid at room temperature (refer to Table 4.1 for the melting and boiling points of ethanol). Ethanol is commonly called alcohol and is the active ingredient of all alcoholic drinks. In addition, because it is a good solvent, it is also used in medicines such as tincture iodine, cough syrups, and many tonics. Ethanol is also soluble in water in all proportions. Consumption of small quantities of dilute ethanol causes drunkenness. Even though this practice is condemned, it is a socially widespread practice. However, intake of even a small quantity of pure ethanol (called absolute alcohol) can be lethal. Also, long-term consumption of alcohol leads to many health problems.

Chemical properties

(i) Reaction with sodium –

$$2Na + 2CH_3CH_2OH \rightarrow 2CH_3CH_2O^-Na^+ + H_2$$
$$\text{(Sodium ethoxide)}$$

Alcohols react with sodium leading to the evolution of hydrogen. With ethanol, the other product is sodium ethoxide.

(ii) Reaction to give unsaturated hydrocarbon: Heating ethanol at 443 K with excess concentrated sulphuric acid results in the dehydration of ethanol to give ethene –

$$CH_3CH_2OH \rightarrow CH_2=CH_2 + H_2O$$

The concentrated sulphuric acid can be regarded as a dehydrating agent which removes water from ethanol.

8.20 ETHANOIC ACID

Physical properties

Ethanoic acid is commonly called acetic acid and belongs to a group of acids called carboxylic acids. 5-8% solution of acetic acid in water is called vinegar and is used widely as a preservative in pickles. The melting point of pure ethanoic acid is 290 K and hence it often freezes during winter in cold climates. This gave rise to its name glacial acetic acid. The groups of organic compounds called carboxylic acids are obviously characterized by a special acidity. However, unlike mineral acids like HCl, which are completely ionized, carboxylic acids are weak acids.

Chemical properties

Esterification reaction: Esters are most commonly formed by reaction of an acid and an alcohol. Ethanoic acid reacts with absolute ethanol in the presence of an acid catalyst to give an ester –

$$CH_3COOH + CH_3CH_2OH \rightarrow CH_3\text{-}CO\text{-}CH_2CH_3$$
(ethanoic acid)   (alcohol)   acid   (Ester)

Esters are sweet-smelling substances. These are used in making perfumes and as flavoring agents. Esters react in the presence of an acid or a base to give back the alcohol and carboxylic acid. This reaction is known as saponification because it is used in the preparation of soap.

Formation of ester
$$CH_3COOC_2H_5 \; C\,H\,OH \rightarrow C_2H_5OH + CH_3COOH$$
$$(NaOH)$$

(i) Reaction with a base: Like mineral acids, ethanoic acid reacts with a base such as sodium hydroxide to give a salt (sodium ethanoate or commonly called sodium acetate) and water:

$NaOH + CH_3COOH \rightarrow CH_3COONa + H_2O$

(ii) Reaction with carbonates and hydrogen carbonates: Ethanoic acid reacts with carbonates and hydrogen carbonates to give rise to a salt, carbon dioxide and water. The salt produced is commonly called sodium acetate.

$2CH_3COOH + Na_2CO_3 \rightarrow 2CH_3COONa + H_2O + CO_2$

$CH_3COOH + NaHCO_3 \rightarrow CH_3COONa + H_2O + CO_2$

COMPARATIVE GUIDE: PROPERTIES OF ETHANOL & ETHANOIC ACID

INSIGHTS INTO THE ESSENTIAL ATOM-ARRANGEMENTS & REACTIVITY PATTERNS OF TWO KEY CARBON COMPOUNDS

NUMBER CORE REACTION TYPE PREFIX/SUFFIX SIDE ICONS REPRESENTATIVE GROUP STRUCTURAL DRAWING BOND STRUCTURE ZOOM-IN

1. PHYSICAL PROPERTIES

ETHANOL (Ethyl Alcohol, C_2H_5OH)
- **State:** Liquid at room temperature.
- **Melting Point:** -114.1°C / **Boiling Point:** 78.4°C
- **Odor:** Characteristic, slightly sweet, burning taste.
- **Density:** 0.789 g/mL.
- **Solubility:** Miscible in water.

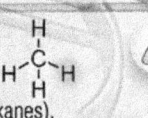

ETHANOIC ACID (Acetic Acid, CH_3COOH)
- **State:** Liquid at room temperature
- **Melting Point:** 16.6°C
- **Boiling Point:** 118.1°C
- **Odor:** Pungent, sour odor (VINEGAR)
- **Density:** 1.049 g/mL
- **Solubility:** Miscible in water

2. SUBSTITUTION
Methane + Cl_2 (diffused sunlight/UV) →
Characteristic of **SATURATED** compounds (Alkanes).
Slow replacement of atoms (e.g., H by Halogen) under light.
Swapping, not breaking the main chain.
(Example: Chlorination of Methane)

2. CHEMICAL PROPERTIES

Combustion
Burns cleanly in air.
Example:
$C_2H_5OH + 3O_2 \rightarrow 2CO_2 + 3H_2O$ + Heat/Light.

Oxidation
Oxidized to Ethanal and then Ethanoic Acid.
Key Reagents: alkaline $KMnO_4$ or acidified $K_2Cr_2O_7$.
$C_2H_5OH \xrightarrow{[KMnO_4]} CH_3CHO \xrightarrow{[O]} CH_3COOH$

Reaction with Sodium
Forms sodium ethoxide and hydrogen gas.
Example: $2C_2H_5OH + 2Na \rightarrow 2C_2H_5ONa + H_2$

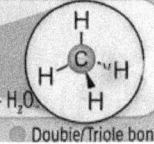

Product of deep alcohol oxidation

Acidic Behavior
Turns blue litmus red.
Reacts with bases.
Example: $CH_3COOH + NaOH \rightarrow CH_3COONa + H_2O$

Reaction with Carbonates
Forms CO_2 and salt.
Example: $2CH_3COOH + Na_2CO_3 \rightarrow 2CH_3COONa + H_2O + CO_2$

Esterification
Reacts with alcohols.

Decarboxylation
Forms methane upon heating with soda-lime.
Example: $CH_3COOH + 2NaOH \rightarrow CH_4 + Na_2CO_3 + H_2O$

SOURCE: CONSOLIDATED CHEMICAL RESOURCE DATA

● Oxygen-red ● Nitrogen ● Halogen ● Double/Triple bond

8.21 SOAPS AND DETERGENTS

Detergents are ammonium or sulphonate salts of long chain carboxylic acids while soaps are sodium salt of carboxylic acid. When calcium or magnesium salts are dissolved in water it becomes hard water and can't be useful for soap due to formation of scum an insoluble substance. However, detergent can be used in hard water because charged ends of detergent do not form scum. The following table shows the difference between the soaps and detergent.

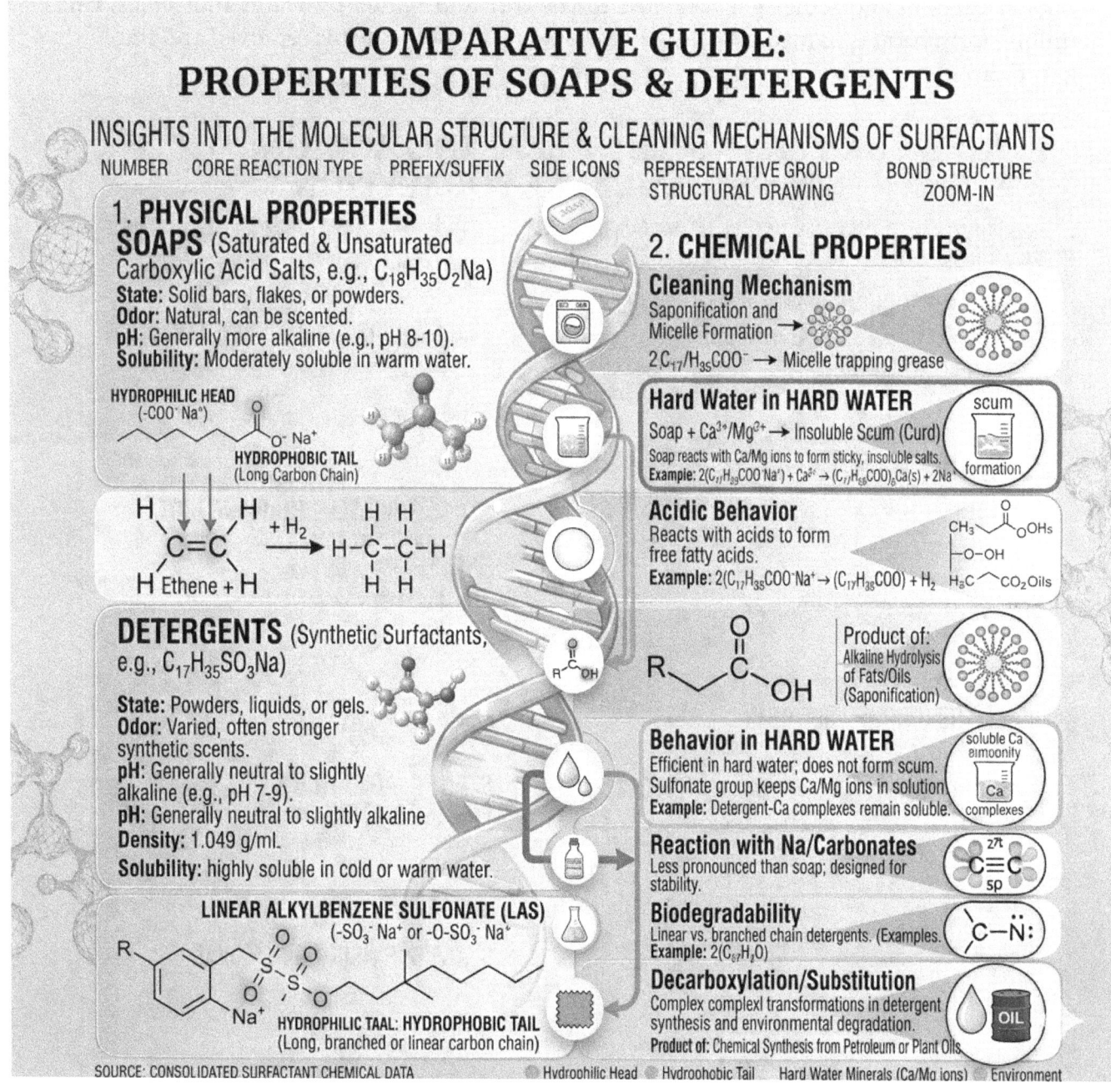

8.22 CLEANING ACTION OF SOAPS

The action of soaps and detergents is based on the presence of both hydrophobic and hydrophilic groups in the molecule and this helps to emulsify the oily dirt and hence its removal. When detergent or soaps are dissolved in water, they produce many molecules which have hydrophobic and hydrophilic ends. Hydrophobic end repels water molecule and gets attached to the dirt while hydrophilic does its opposite. In this process all hydrophobic parts of molecules get attached to the dirt and pulled by water molecules, this unique formation is named *micelle*. When we wash out water dirt is removed and the cloth gets cleaned.

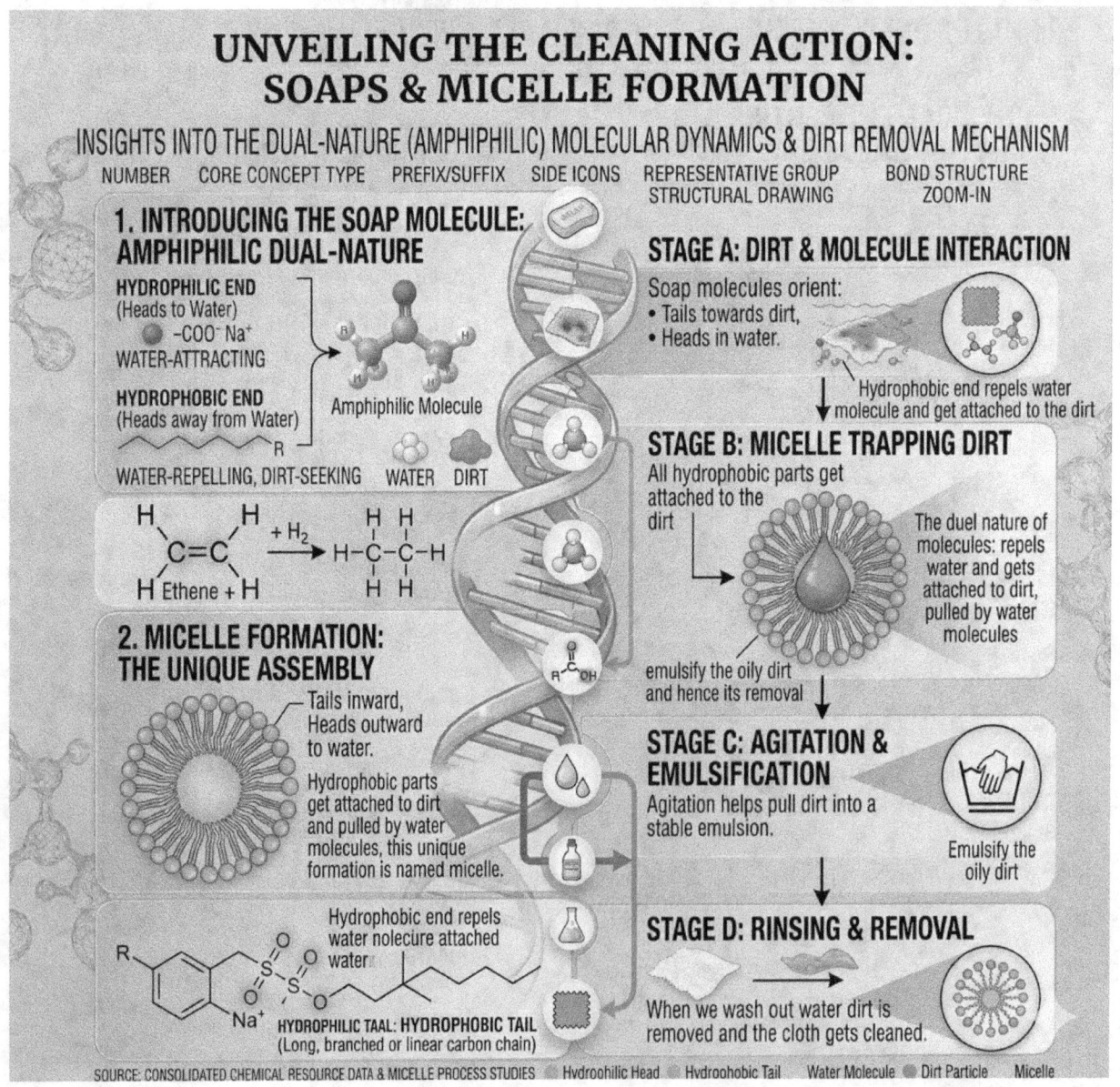

CHAPTER 8: DEEP DIVE CHALLENGES

Q1. The % of carbon in earth crust is?

(a) 0.002

(b) 0.02

(c) 20

(d) 0.2

Q2. Carbon has valency?

(a) 1

(b) 2

(c) 3

(d) 4

Q3. Catenation is found in?

(a) C

(b) O

(c) F

(d) None of these

Q4. Bonding in carbon is?

(a) covalent

(b) ionic

(c) c.co-ordinate

(d) none of these

Q5. A saturated hydrocarbon has?

(a) Single bond

(b) Double bond

(c) Triple bond

(d) None of these

Q6. Formula of benzene is?

(a) C_2H_6
(b) C_6H_6
(c) C_5H_6
(d) None of these

Q7. Homologous series differ by?

(a) CH_2 unit
(b) 14 amu by mass
(c) Both a and b
(d) None of these

Q8. $CH_3CH_2CH_2OH$ is?

(a) alcohol
(b) ketone
(c) carboxylic acid
(d) None of these

Q9. In oxidation reaction the oxidising agent is?

(a) Alkaline $KMnO_4$
(b) Acidified $K_2Cr_2O_7$
(c) Both a and b
(d) None of these

Q10. In addition reaction the catalyst used is?

(a) H_2
(b) Acidified $K_2Cr_2O_7$
(c) nickel
(d) None of these

Q11. The conversion of ethanol to ethanoic acid is?

(a) Oxidation reaction
(b) Addition reaction
(c) Substitution reaction
(d) None of these

Q12. Which is commonly known as alcohol?

(a) ethanol
(b) ethanoic acid
(c) methanol
(d) None of these

Q13. Vinegar is formed from?

(a) ethanol
(b) ethanoic acid
(c) methanol
(d) None of these

Q14. In esterification the catalyst used is?

(a) base
(b) acid
(c) salt
(d) None of these

Q15. Saponification is the reverse process of?

(a) carbocation
(b) esterification
(c) carbocation
(d) None of these

Q16. Ionic end of soap dissolves in?

(a) water
(b) oil
(c) acid
(d) base

Q17. Shampoos are?

(a) soap
(b) detergent
(c) salts
(d) acids

Q.18 which of the following can go under addition reaction?

(a) C_2H_6
(b) C_6H_6

(c) C5H6

(d) None of these

Q19. CH3COOCH3 is?

(a) soap

(b) acid

(c) base

(d) ester

Q20. Which of the following causes hardness of water?

(a) Calcium salt

(b) Magnesium salt

(c) Both a and b

(d) None of these

Match the following:

Q21.

A	B
C-60	Saturated
Ethene	Allotropes
Methane	Unsaturated
Carbon	Catenation

Q22.

A	B
-OH	Carboxylic Acid
-CHO	Ketone
-CO-	Aldehyde
-COOH	Alcohol

Q23.

A	B
Nickel	Substitution Reaction
Sun light	Addition Reaction
Alkaline KMnO$_4$	Oxidation Reaction
Hot Conc. H$_2$SO$_4$	Formation of ethene

Short Answer Questions

Q24. What are the versatile properties of carbon?

Q25. Write down differences between saturated and unsaturated hydrocarbons?

Q26. What is homologous series?

Q27. What is addition reaction?

Q28. What is substitution reaction?

Q29. Write down physical properties of ethanol?

Q30. Write down physical properties of ethanoic acid?

Q31. Write down chemical properties of ethanol?

Q32. Write down chemical properties of ethanoic acid?

Long Answer Questions:

Q33. Draw the structures for the following compounds.
    (i) Ethanoic acid   (ii) Bromopentane*
    (iii) Butanone   (iv Hexanal.
    *Are structural isomers possible for bromopentane?

Q34. Explain the cleaning action of soap?

Q35. Explain the phenomenon of esterification and saponification in detail with examples.

Solution

1.b 2.d 3.a 4.a 5.a 6.b 7.c 8.a 9.c 10.a 11.a 12.a 13.b 14.b 15.b 16.a 17.b 18.b 19.d 20.c

Short Answer Hints
- Q24 (Versatile Carbon): Think about carbon's ability to form long chains (Catenation) and its "four hands" (Tetravalency) that allow it to bond with many different elements.
- Q25 (Saturated vs. Unsaturated): It's all about the "handshakes." Do the carbon atoms share a single bond, or are there double or triple bonds between them?
- Q26 (Homologous Series): This is a "family" of compounds. They have the same functional group and similar chemical properties, but each member differs from the next by a specific group—think -CH2-.
- Q27 (Addition Reaction): Think of "opening a gate." This happens in unsaturated compounds where a double bond breaks to let new atoms in without removing any old ones.
- Q28 (Substitution Reaction): Think of "musical chairs." This happens in saturated hydrocarbons where one atom (usually Hydrogen) is pushed out and replaced by another (like Chlorine).
- Q29 (Physical Ethanol): Think of its state at room temperature, its distinct smell (like spirit), and whether it mixes with water.
- Q30 (Physical Ethanoic Acid): Think of Vinegar. It has a pungent smell and can freeze

into a "glacier-like" solid in cold climates.
- Q31 (Chemical Ethanol): Focus on its reaction with Sodium metal (releasing H2 gas) and its Dehydration to form Ethene using hot concentrated H2SO4.
- Q32 (Chemical Ethanoic Acid): Focus on Esterification (making a fruity smell) and its reaction with carbonates to release CO2 (fizzing).

Long Answer Hints

Q33: Structures and Isomers
- (i) Ethanoic acid: Needs a 2-carbon chain ending in a -COOH group.
- (ii) Bromopentane: A 5-carbon chain with a Br atom. (Isomer Hint: Can the Br be placed on the first, second, or third carbon?)
- (iii) Butanone: A 4-carbon chain with a double-bonded oxygen (C=O) in the middle.
- (iv) Hexanal: A 6-carbon chain ending in an aldehyde group (-CHO).

Q34: Cleaning Action of Soap
- Think of a soap molecule as a "tadpole." The Hydrophilic head loves water, but the Hydrophobic tail hates water and loves oil.
- How do these tails trap the dirt in the center to form a ball-like structure called a Micelle?

Q35: Esterification vs. Saponification
- Esterification: Think: Acid + Alcohol $\rightarrow$ Sweet-smelling Ester. It's like making a perfume.
- Saponification: It's the "Reverse" of esterification. When you treat an Ester with a Base (NaOH), it breaks back down to give you Soap and Alcohol.

CHAPTER 9

PERIODIC CLASSIFICATION OF ELEMENTS

9.1 INTRODUCTION

Why classification is important?

Classification provides us an easy way to understand the properties and access to the particular element. We can take a simple example; in a mall the items pertaining to same purposes are kept one side. If all items present in the mall are spread and mixed, what will happen? It will create a huge disturbance to customer as well as managing team. If it is classified, it is quite easy for selling as well as purchasing. If all elements are arranged according to some particular properties, it will be quite easier to select a particular element for a particular purpose. Suppose any engineer wants to form the body part of an aero plane. Then first, he will be attempted to ignore liquids and gases. Then he will try to choose light metals which have high tensile strength. Further he may be interested in other properties too. All these comparisons can be done only when all elements are arranged in a particular way. But before 200 years when classification starts, it was a great quest, how to start it? There came thousands of ways and tables some were discarded; some were improved and gave a platform to next one. In this chapter we will see only some of the classification which proved to the base for modern periodic classification of elements. In this chapter our strategy will be to have a look at the basic rule for a particular classification then its merits and demerits. We will also deal with modern periodic table in details.

9.2 DOBEREINER'S TRIADS

In the year 1817, Johann Wolfgang Döbereiner, a German chemist, arranged elements in the group of three named them as triads. According to dobereiner, when these three elements are arranged in increasing order of their atomic weight the middle weight is average of remaining two.

Demerits

Dobereiner could arrange only three groups.

9.3 NEWLAND'S OCTAVES

Doberiener's triads encouraged many researchers to search periodicity in elements according to the increasing order of mass. Newlands took the hydrogen as lightest element and thorium as 56th and last element. On arranging these elements in increasing order of their atomic mass, Newlands found that every eighth element shows similar properties. He proposed a word "the octaves law" for this periodicity.

Demerits of Newlands octaves

1. Newlands law of octaves was applicable to only light elements. It was well up to calcium element.
2. Newland announces that there are only 56 elements in the nature and no further elements will be discovered. But soon it was realized that it is not true.
3. Some elements were put in same slot like cobalt and nickel and both of these were in the column of fluorine which has different properties. Iron which resembles properties same as cobalt was kept far away from it.

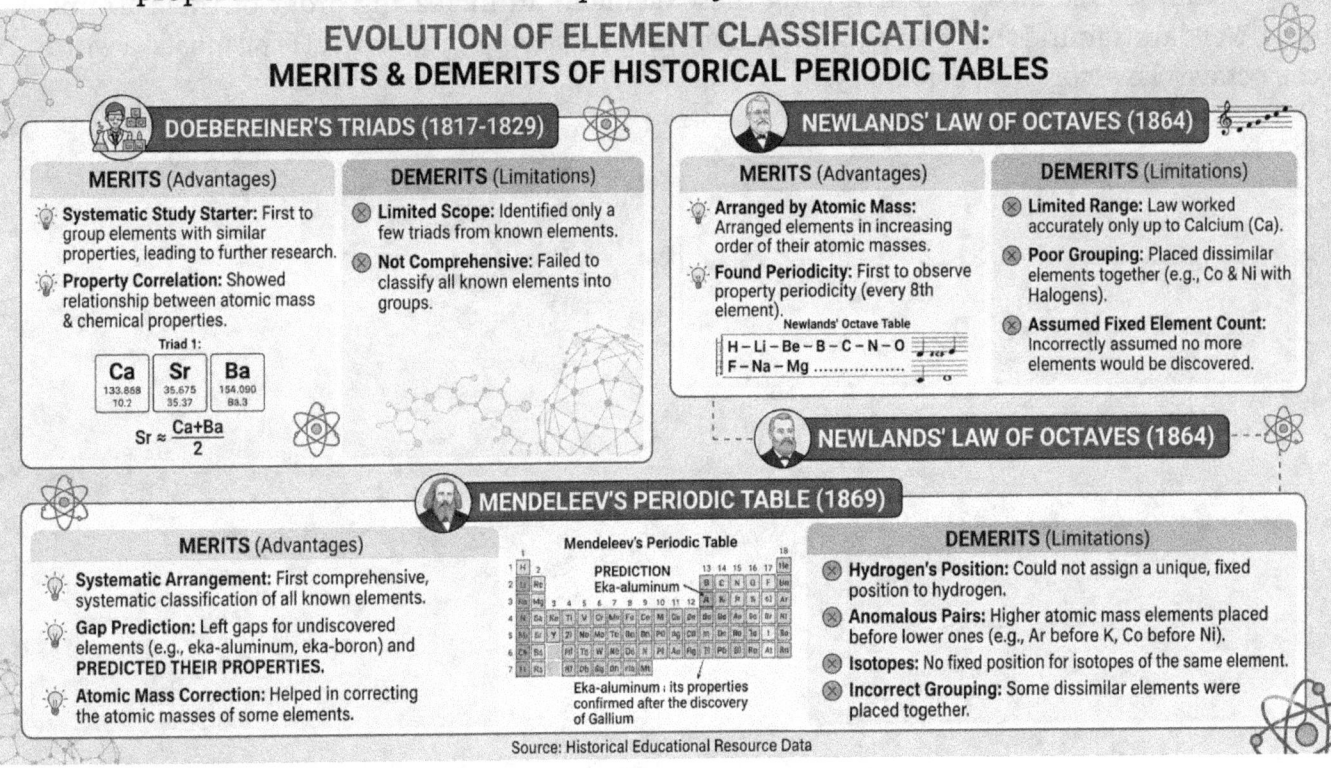

9.4 MENDELEEV'S PERIODIC TABLE

When Mendeleev started preparing the periodic table there were 63 elements. Mendeleev investigated the periodic function of masses as well as physical and chemical properties. He formed oxides and hydrides of each element and kept similar in a same group. Then he stated the law of periodic table which states that which states that

'The properties of elements are the periodic function of their atomic masses'.

9.5 ACHIEVEMENTS OF MENDELEEV'S PERIODIC TABLE:

Mendeleev's left some gaps in the periodic table. He took it as a merit and explained that some elements will be discovered in the future and will be kept at these positions. These elements were scandium, gallium, germanium and were respectively named as **Eka–boron, Eka–aluminium and Eka–silicon**. Here eka (a Sanskrit word) means one. Eka word was prefix to the element preceding in the same group. Predictions of properties were almost same as these elements have. His periodic table was so flexible that when inert gases were discovered they got their places easily.

9.6 LIMITATIONS OF MENDELEEV'S PERIODIC TABLE

Mendeleev was *unable to give a fix position for hydrogen*. His periodic table was based on physical and chemical properties and hydrogen has these properties similar to metal as well as nonmetals.

As periodic table was arranged in according to the increasing order of their atomic mass there was *no any place for isotopes*.

9.7 THE MODERN PERIODIC TABLE

This periodic table was prepared by Henry Moseley in 1913. In this table Moseley took the atomic no. as the as a fundamental property rather than atomic mass. This concept solved many limitations of Mendeleev's periodic table. This law can be stated as follows:

'Properties of elements are a periodic function of their atomic number.'

9.8 POSITION OF ELEMENTS IN MODERN PERIODIC TABLE

In modern periodic table, there are 7 horizontal rows known as periods and 18 vertical columns known as groups. The elements having same chemical and physical properties are arranged in the same group. However, in periods a
gradation or slight change in physical and chemical properties is observed.

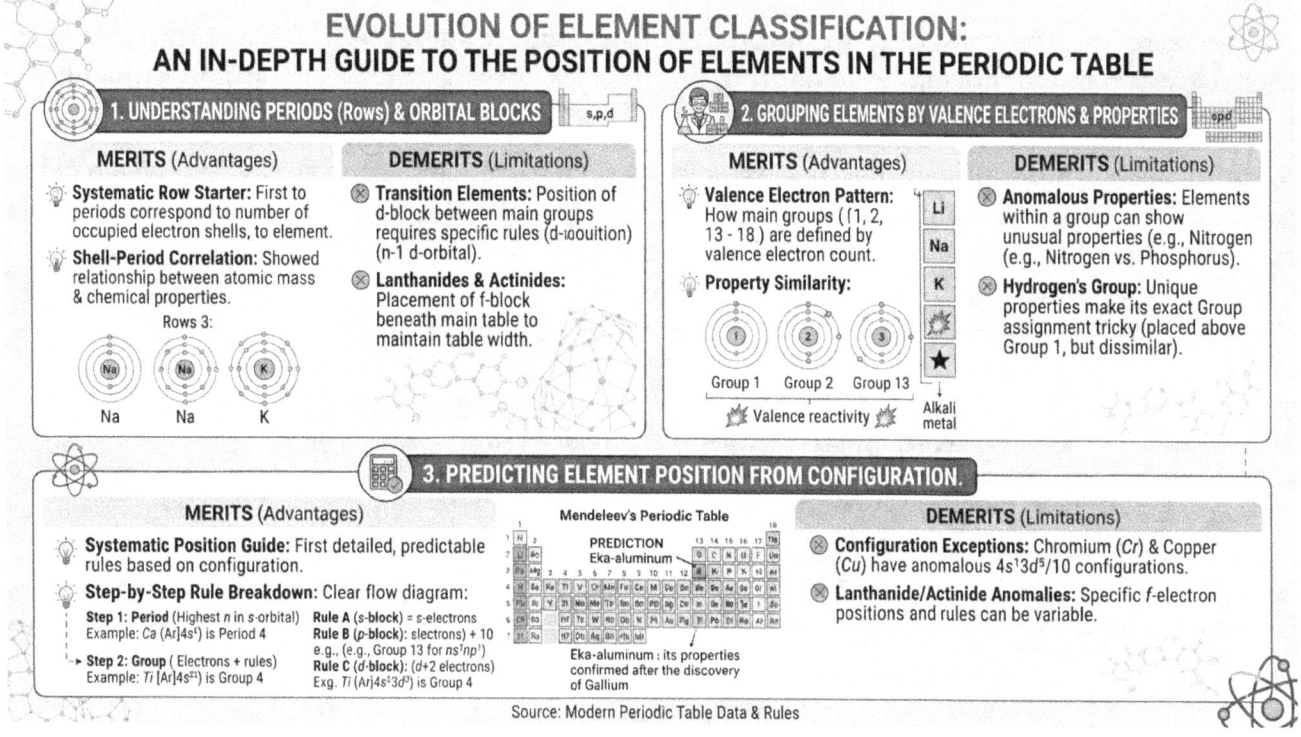

9.9 TRENDS IN THE MODERN PERIODIC TABLE

Valency: in a group the valency remains unchanged as there is same no. of electrons in outer most shell of atom. When we move from left to right in a period the valency first increases and then decreases. We have already seen how to calculate valency in "basic chemistry 1" if we take the second period of modern periodic table the elements Li, Be, B, C, N, O, F, and Ne we get the valencies as 1, 2, 3, 4, 3, 2, 1, 0 respectively.

Atomic size: in group as we go from top to bottom, due to increase in no. of orbital atomic size increases. When we move in periods from left to right the no. of orbitals remain same but no. of protons increases which attracts electrons and size shrinks i.e. atomic size decreases.

Metallic and nonmetallic properties: as we have seen in above section that when we go from top to bottom in groups the atomic size Increases and hence the electrons in the outer most shells are freer to move, this contributes to metallic properties.
Therefore, ongoing from top to bottom in groups metallic properties increase. Similar logic can be applied to see the metallic properties decreases on going from left to right in periods.

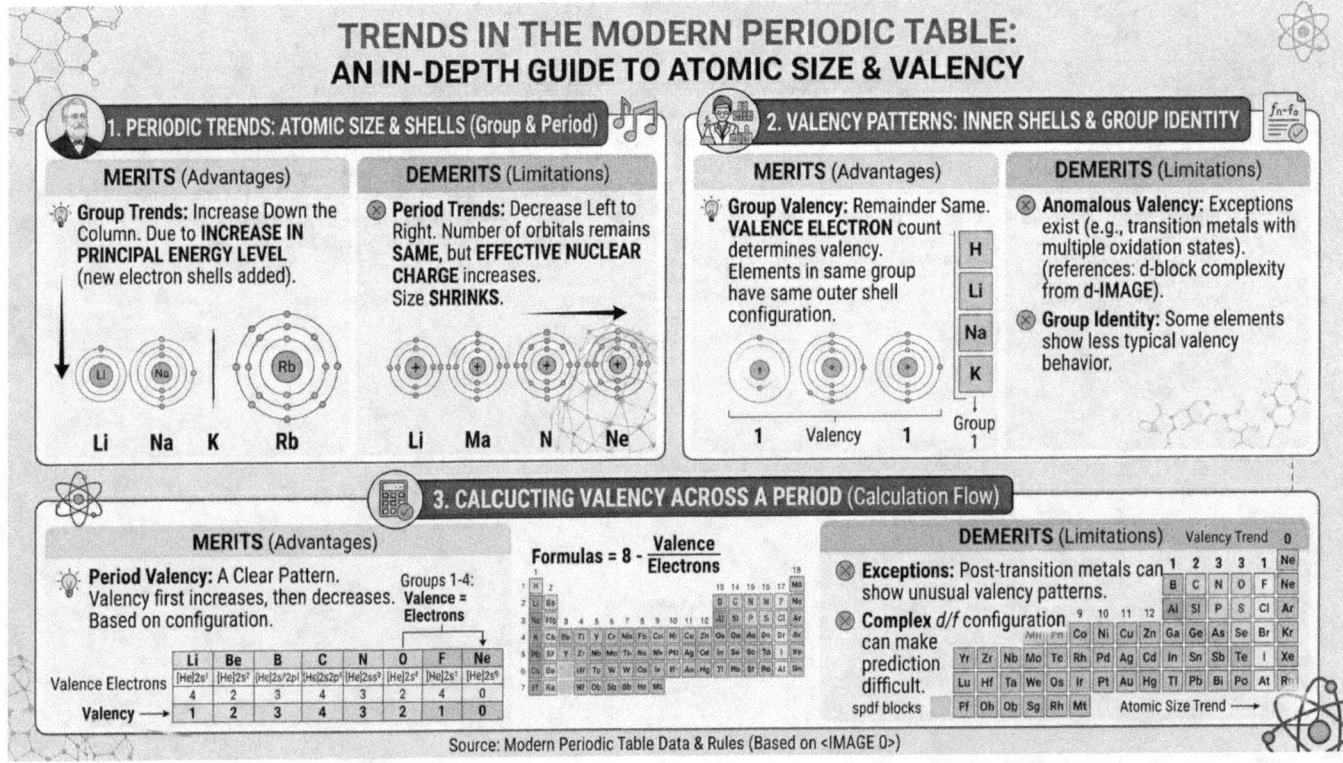

Source: Modern Periodic Table Data & Rules (Based on <IMAGE 0>)

CHAPTER 9: DEEP DIVE CHALLENGES

Q1. Doberiener arranged the elements in the form of?

(a) triads
(b) octaves
(c) tetrads
(d) none

Q2. Newlands arranged how many elements?

(a) 56
(b) 63
(c) 114
(d) 118

Q3. Mendeleev left gape for which element in his periodic table?

(a) scandium
(b) gallium
(c) germenium
(d) all of these

Q4. Mendeleev used basic concept of?

(a) Physical properties
(b) Chemical properties
(c) Atomic masses
(d) all of these

Q5. Who prepared modern periodic table?

(a) newland
(b) doberiener
(c) mendeleev
(d) moselley

Q6. How many periods are there in modern periodic table?

(a) 18
(b) 7
(c) 20
(d) None of these

Q7. What is the valency of magnesium?

(a) 1
(b) 2
(c) 3
(d) 4

Q8. How many groups are there in modern periodic table?

(a) 18
(b) 7
(c) 20
(d) None of these

Q9. In periods going right from left, the valency?

(a) increases
(b) decreases
(c) first increases then decreases
(d) None of these

Q10. In groups going from top to bottom, the valency?

(a) increases
(b) decreases
(c) first increases then decreases
(d) remain same.

Q11. In periods going right from left, atomic size?

(a) increases
(b) decreases
(c) first increases then decreases
(d) None of these

Q12. In groups going from top to bottom, the atomic size?

(a) increases
(b) decreases

(c) first increases then decreases

(d) remain same.

Q13. In groups going from top to bottom, the metallic properties?

(a) increases

(b) decreases

(c) first decreases then increases

(d) remain same.

Q14. In periods going right from left, the non mettalic properties?

(a) increases

(b) decreases

(c) first increases then decreases

(d) None of these

Match the followings

Q15.

A	B
Law of octaves	Moselley
Triads	Mendeleev
Gaps in table	Newlands
Place for isotops	Doberiener

Q16.

A	B
56 elements	Moselley
63 elements	Mendeleev
9 elements	Newlands
118 elements	Doberiener

Long answers type questions:

Q17. Explain the trends in the periodic table for

(a) Valency

(b) Atomic size

(c) Mettalic and non metallic properties

Q18. Which element has

(a) Two shells, both of which are completely filled with electrons?

(b) The electronic configuration 2, 8, 2?

(c) A total of three shells, with four electrons in its valence shell?

(d) A total of two shells, with three electrons in its valence shell?

(e) twice as many electrons in its second shell as in its first shell?

Q19.
  (a) What property do all elements in the same column of the Periodic Table as boron have in common?
  (b) What property does all elements in the same column of the Periodic Table as Fluorine has in common?

Q20. The position of three elements A, B and C in the Periodic Table are shown below –

Group 16	Group 17
-	-
-	A
-	-
B	C

(a) State whether A is a metal or non-metal.
(b) State whether C is more reactive or less reactive than A.
(c) Will C be larger or smaller in size than B?
(d) Which type of ion, cation or anion, will be formed by element A?

Solution

1.a 2.a 3.d 4.d 5.d 6.b 7.b 8.a 9.c 10.d 11.b 12.a 13.a 14.a

Long Answer Hints
Q17: Trends in the Periodic Table
Think of the periodic table as a grid where properties change as you move Across a Period (left to right) and Down a Group (top to bottom).
- (a) Valency: In a period, it increases and then decreases (1, 2, 3, 4, 3, 2, 1, 0). In a group, it stays exactly the same. Why? (Hint: Think of the number of electrons in the outermost shell).
- (b) Atomic Size: As you move down, you add new "floors" (shells) to the atom. As you move across, the center (nucleus) gets "stronger" and pulls the electrons in tighter. Does this make the atom bigger or smaller?
- (c) Metallic and Non-Metallic: Metals like to lose electrons, while non-metals like to gain them. Is it easier to lose an electron when the nucleus is far away (bottom left) or when it's being pulled very tightly (top right)?

Q18: Identifying Elements
- (a) Both shells full means Shell 1 has 2 and Shell 2 has 8. (Hint: It's a Noble Gas).
- (b) Add the numbers 2 + 8 + 2. Which element has this atomic number? (Hint: It's in Group 2).
- (c) Three shells means it's in Period 3. Four valence electrons mean it's in Group 14. (Hint: Think of the "computer chip" element).

- (d) Two shells mean Period 2. Three valence electrons mean Group 13.
- (e) If the first shell always has 2, then the second shell must have 4. What is the element with atomic number 6?

Q19: Group Characteristics
- (a) Boron Column (Group 13): Look at the valence electrons. If Boron has 3, how many do the elements below it have?
- (b) Fluorine Column (Group 17): These are the Halogens. They all need exactly *one* more electron to be "happy" (stable). What does that say about their valency and reactivity?

Q20: Periodic Table Logic
- (a) Metal or Non-metal: Look at the Group numbers. Elements in Group 16 and 17 are on the far-right side of the periodic table. Do elements on the right typically like to gain electrons or lose them?
- (b) Reactivity of C vs. A: Both are in Group 17 (the Halogens). In this specific group, the "hunger" for electrons is strongest at the top because the nucleus is closer to the incoming electron. As you go down the group, does the ability to attract an electron increase or decrease?
- (c) Size of C vs. B: Both are in the same period (row). As you move from left to right across a period, the "positive pull" of the nucleus increases while the number of shells stays the same. Does this extra pull make the atom expand or shrink?
- (d) Type of Ion: Element A is in Group 17, meaning it has 7 valence electrons. Is it easier for it to lose all 7 electrons or just "grab" 1 more to become stable? If it gains a negative electron, what charge does it get?

APPENDIX A: STANDARD REDUCTION POTENTIALS (25°C)

Essential for predicting the feasibility of displacement reactions and electrochemical cell potential.

Element/Ion	Half-Reaction	E0 (Volts)
Lithium	$Li^+ + e^- \rightarrow Li$	-3.04
Potassium	$K^+ + e^- \rightarrow K$	-2.93
Calcium	$Ca^{2+} + 2e^- \rightarrow Ca$	-2.87
Sodium	$Na^+ + e^- \rightarrow Na$	-2.71
Magnesium	$Mg^{2+} + 2e^- \rightarrow Mg$	-2.37
Aluminium	$Al^{3+} + 3e^- \rightarrow Al$	-1.66
Zinc	$Zn^{2+} + 2e^- \rightarrow Zn$	-0.76
Iron	$Fe^{2+} + 2e^- \rightarrow Fe$	-0.44
Hydrogen	$2H^+ + 2e^- \rightarrow H_2$	0.00
Copper	$Cu^{2+} + 2e^- \rightarrow Cu$	+0.34
Silver	$Ag^+ + e^- \rightarrow Ag$	+0.80

Gold	$Au^{3+} + 3e^- \rightarrow Au$	+1.50
Fluorine	$F^{2+} 2e- \rightarrow 2F^-$	+2.87

APPENDIX B: IUPAC FUNCTIONAL GROUP PRIORITY TABLE

Use this table to determine the principal functional group for naming complex organic compounds (Chapter 8).

Priority	Functional Group	Formula	Prefix	Suffix
1	Carboxylic Acid	-COOH	Carboxy-	-oic acid
2	Ester	-COOR	Alkoxycarbonyl-	-oate
3	Amide	$-CONH_2$	Carbamoyl-	-amide
4	Nitrile	-CN	Cyano-	-nitrile
5	Aldehyde	-CHO	Formyl-	-al
6	Ketone	>C=O	Oxo-	-one
7	Alcohol	-OH	Hydroxy-	-ol
8	Amine	$-NH_2$	Amino-	-amine
9	Alkene	>C=C<	---	-ene
10	Alkyne	$-C \equiv C-$	---	-yne

APPENDIX C: COMMON ISOTOPES AND THEIR APPLICATIONS

Expanding on the foundational concepts in Chapter 1.

Isotope	Symbol	Application/Use
Carbon-14	^{14}C	Carbon dating of fossils and organic remains.
Cobalt-60	^{60}Co	Radiation therapy to treat cancer and sterilizing medical equipment.
Iodine-131	^{131}I	Diagnosis and treatment of thyroid gland disorders.
Uranium-235	^{235}U	Fuel for nuclear reactors and power generation.
Phosphorus-32	^{32}P	Agricultural research (tracking fertilizer uptake in plants).
Americium-241	^{241}A	Used in ionizing smoke detectors.

APPENDIX D: GENERAL SOLUBILITY RULES FOR AQUEOUS SOLUTIONS

A vital tool for predicting precipitates in chemical reactions (Chapter 5).

Soluble Compounds (No Precipitate)	Exceptions (Forms Precipitate)
All Group 1 Metals (Li^+, Na^+, K^+, etc.)	None
All Ammonium salts (NH_4^+)	None
All Nitrates (NO_3^-) and Acetates	None
Chlorides, Bromides, Iodides	Ag^+, Pb^{2+}, Hg_2^{2+}
Sulfates (SO_4^{2-})	Ca^{2+}, Sr^{2+}, Ba^{2+}, Pb^{2+}

Insoluble Compounds (Forms Precipitate)	Exceptions (Soluble)
Carbonates & Phosphates	Group 1 Metals and NH_4^+
Hydroxides (OH^-)	Group 1 Metals, Ba^{2+}, and Sr^{2+}
Sulfides (S^{2-})	Group 1, Group 2 Metals, and NH_4^+

NOTES

EPILOGUE

The conclusion of a book is rarely the conclusion of a subject. In the world of science and technology, every answer we uncover simply provides a better lens through which to view the next question.

Throughout these chapters, we have dismantled complex theories and rebuilt them into a foundation you can stand upon. My hope is that you no longer view these concepts as static lines in a textbook, but as a dynamic language that describes the world around us.

As you move forward whether toward an examination, a career in research, or personal mastery carry with you the spirit of persistent inquiry. The most successful students are not those who know all the answers, but those who never stop refining their questions. The circuit is now complete, but the power is in your hands.

AFTERWORD

Writing this book has been a journey of both reflection and reinvention. After fifteen years of teaching and authoring works in the academic sphere, I found myself returning to the roots of why I began this path in the first place: the simple, profound desire to make knowledge accessible.

This volume was drafted not in a bustling city office, but in the quiet surroundings of my native village, Nipaniya. It was here, away from the noise, that I could truly focus on distilling these complex ideas into their most useful forms. This book represents more than just information; it represents a commitment to the idea that high-quality education should be available to everyone, regardless of where they are starting from.

Thank you for being part of this mission. Every reader who finds clarity in these pages brings me one step closer to the goal of bridging the gap between traditional learning and the digital future.

ACKNOWLEDGEMENT

No book is the product of a single mind. It is a tapestry woven from the support, patience, and inspiration of many.

First, I must express my deepest gratitude to my students. Your endless questions and pursuit of excellence have not only challenged me as an educator but have also shaped the clarity and structure of this work. You are the reason I write.

I am profoundly thankful to my family for their unwavering support, especially during the long hours of research and writing. To my nephew, Shobhit, and my nieces, Palak and Khushi your energy and aspirations remind me daily of the importance of passing on knowledge to the next generation.

Finally, I would like to thank the academic community and my colleagues who have provided a sounding board for these ideas over the last fifteen years. This work is as much yours as it is mine.

ABOUT THE AUTHOR

Umesh Kumar

Umesh Kumar is a dedicated academician, educator, and prolific author with a deep-seated passion for making complex scientific concepts accessible to students. With over a decade of experience in higher education, he currently serves as an Assistant Professor of Electronic Science and has held faculty positions at prestigious institutions, including the University of Delhi and Delhi Skill and Entrepreneurship University (DSEU). A lifelong scholar, Umesh holds a B.Sc. (Honours) in Electronics from the University of Delhi and an M.Sc. in Electronic Science from Jamia Millia Islamia. His academic excellence is highlighted by qualifying the UGC NET in Electronic Science four times, reflecting his mastery of the subject and commitment to educational standards. Beyond the classroom, he serves as a qualified Academic Counselor at IGNOU, providing guidance to the next generation of distance learners. As an author, Umesh has made a significant impact in the digital publishing space, with a catalog of over hundred books published on Amazon Kindle Direct Publishing (KDP). His work focuses primarily on academic study guides, fast-revision series, and competitive exam preparation, helping thousands of students navigate rigorous subjects like Chemistry, Physics, and Mathematics with clarity and confidence.Umesh continues to inspire students to look beyond rote learning and embrace the true beauty of scientific inquiry

CHEMISTRY: THE "MASTERY & WORKBOOK"

The Complete Chemistry Collection: From Fundamentals to Advanced Theory
Welcome to the definitive library for modern chemistry students. Authored by an established academic professional with years of experience in Electronic Science and education, this series bridges the gap between classroom lectures and independent mastery.With over 100 titles in the broader academic catalog, these chemistry-specific volumes represent a curated selection of best-in-class study materials. Each title utilizes a structured pedagogical ap
proach, breaking down atomic models, thermodynamics, and molecular structures into digestible, logical modules.Why students choose this series:
Clarity & Precision: Complex theories explained in plain English.Comprehensive Coverage: A one-stop shop for high school, undergraduate, and competitive test prep. Proven Results: Optimized for quick retention and long-term conceptual clarity.Empower your studies today. Scroll down to browse our specialized guides and survival manuals.

Chemical Reactions

This book is genuinely written for grasping the fundamental concept of chemistry. It is aimed to the secondary level students. It can serve as a reference for a particular topic. It is also useful for various competitions.
This book deals with the types of chemical reactions, corrosion, and rancidity. It explores the acid bases and salt around us. There is an additional chapter on metals and non metals, their properties, extraction and uses in daily life.

BOOKS BY THIS AUTHOR

The Ap Chem Survival Guide: Mastering The 10 Most Likely Frq Topics

Stop Guessing. Start Scoring.
The AP Chemistry exam isn't just a test of what you know—it's a test of how well you can justify your answers. Every year, thousands of students lose points not because they got the math wrong, but because they didn't use the specific "scientific claim" language the College Board demands.The AP Chem Survival Guide is built to solve that exact problem. Instead of re-reading a 600-page textbook, this guide surgically targets the 10 Most Likely FRQ Topics—the "Sticking Points" that consistently appear on the exam and determine the difference between a 3 and a 5.
What's Inside the Survival Guide:
The 10-Topic Blueprint: From Atomic Structure & Periodicity to Applications of Thermodynamics, we cover the heavy hitters including Kinetics, Equilibrium, and Acids & Bases.The CER Framework Mastery: Learn to write perfect Free Response answers using the Claim-Evidence-Reasoning model. We show you exactly how to earn the "Justification Point" every time.The "Unit Trap" Warning System: Avoid the classic mistakes that sink scores, like the "J vs. kJ"
thermodynamics trap and the "Shielding Myth" in periodic trends.The Lab Bench Series: A dedicated deep dive into experimental design, error analysis, and titration strategies—skills that are essential for the long-form FRQs.Real-World Application: Master complex concepts like Photoelectron Spectroscopy (PES), VSEPR Theory, and Galvanic Cells through clear, concise breakdowns.
Why This Guide? Most prep books are too long; most cheat sheets are too short. The AP Chem Survival Guide is the "Goldilocks" of prep:
Strategic: Built around the 2026 exam standards.
Efficient: Designed for the "Peak Panic" window when you need results fast.
Results-Driven: Focuses on the scoring rubrics used by actual AP graders.
Don't just survive the exam—master it. Add The AP Chem Survival Guide to your study arsenal today

www.ingramcontent.com/pod-product-compliance
Lightning Source LLC
Chambersburg PA
CBHW081112170526
45165CB00008B/2427